Petroleum Engineering

Editor-in-Chief

Gbenga Oluyemi, Robert Gordon University, Aberdeen, Aberdeenshire, UK

Series Editors

Amirmasoud Kalantari-Dahaghi, Department of Petroleum Engineering, West Virginia University, Morgantown, WV, USA

Alireza Shahkarami, Department of Engineering, Saint Francis University, Loretto, PA, USA

Martin Fernø, Department of Physics and Technology, University of Bergen, Bergen, Norway

The Springer series in Petroleum Engineering promotes and expedites the dissemination of new research results and tutorial views in the field of exploration and production. The series contains monographs, lecture notes, and edited volumes. The subject focus is on upstream petroleum engineering, and coverage extends to all theoretical and applied aspects of the field. Material on traditional drilling and more modern methods such as fracking is of interest, as are topics including but not limited to:

- Exploration
- Formation evaluation (well logging)
- Drilling
- Economics
- Reservoir simulation
- Reservoir engineering
- Well engineering
- Artificial lift systems
- Facilities engineering

Contributions to the series can be made by submitting a proposal to the responsible publisher, Anthony Doyle at anthony.doyle@springer.com or the Academic Series Editor, Dr. Gbenga Oluyemi g.f.oluyemi@rgu.ac.uk.

Tarek Al-Arbi Omar Ganat

Modern Pressure Transient Analysis of Petroleum Reservoirs

A Practical View

 Springer

Tarek Al-Arbi Omar Ganat
Department of Petroleum and Chemical
Engineering
Sultan Qaboos University
Muscat, Oman

ISSN 2366-2646   ISSN 2366-2654 (electronic)
Petroleum Engineering
ISBN 978-3-031-28891-3   ISBN 978-3-031-28889-0 (eBook)
https://doi.org/10.1007/978-3-031-28889-0

This Springer imprint is published by the registered company Springer Nature Switzerland AG
The registered company address is: Gewerbestrasse 11, 6330 Cham, Switzerland

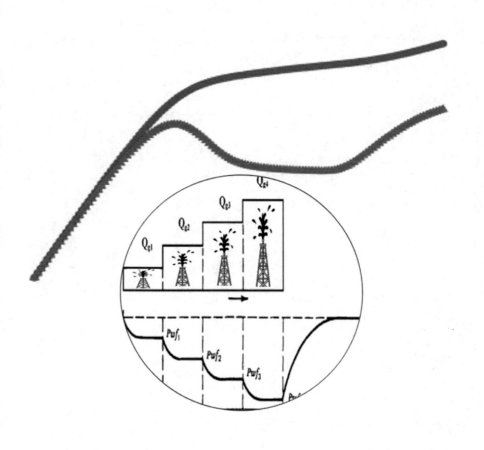

This book is dedicated to my beloved parents, my siblings, my wife Basma, my children Mohamed, Heba, Abdulrahman, and my young hero Abdul Malik. Without their support and inspiration, this work will not be produced

Preface

Well test analytical procedures have developed dramatically during the past decade. With the advent of high-resolution pressure measurements and sophisticated processors, information from well testing has become more reliable and relevant. The novel interpretation approaches, which use the pressure derivative, enhance the distinctive subsurface reservoir characterization. The variety of theoretical options available to the interpretation experts is improving all the time as well test data diagnosis improves. Today's well test interpretation computer programs can analyze pressure transient test results for a wide variety of complicated well and reservoir configurations.

This book covers all areas of well test design and analysis, for well testing experts' engineers and students. Many examples of well test interpretation models are discussed in detail, and their use in field observations is verified. The practical analysis of well test results is fully explained. If the recorded test data deviates from the theory because of operational conditions, golden rules are devised for proper interpretation. The fundamental well test analysis technique and accompanying theories have been extensively explored in the literature, and they are clearly explained here. From this book, well test engineers not only discover solutions for the various problems that arise throughout the evaluation, but they will gain a better understanding of the fundamental process involved, as well as the significance and constraints of the results.

Considering the well testing analysis which is more important to reservoir engineering study, this book makes an effort to produce a thorough reference to cover the topic. The objective of this book is not to duplicate or repeat the conventional well testing courses and/or references. On the contrary, it focuses on the significance of well test qualities and interpretation for reservoir characterization studies.

Hope that everyone enjoys reading this book and finds it useful in his well testing work. Any scientific subject in this book reflects my best understanding. However, as technical individuals, we must always explore different perspectives.

Muscat, Oman   Tarek Al-Arbi Omar Ganat

Overview of Chapters

This book presents nine chapters of technical knowledge about pressure transient analysis. All of the chapters cover different subjects and provide the reader with a complete understanding of the well testing methodology and analysis, enabling well test engineers to comprehend the interpretation process together with the procedures required for reservoir interpretation.

The book will highlight the practical use of contemporary techniques in well test analysis, with a focus on the pressure derivative in particular. Flow tests, buildup tests with or without phase redistribution, multi-rate testing, interference tests, and pulse tests will all be analyzed using various pressure analysis techniques. Pressure test interpretation under multiphase flow regimes, hydraulically fractured wells, naturally fractured reservoirs, and horizontal wells will be covered in detail. Based on the various forms of well tests and the amount of the obtained well test data, the chapters were written in a simple way to provide a simple reference and guide.

Additionally, this book provides solutions to the exercises presented in various sections to highlight the procedures addressed in every chapter, helping well test engineers to better understand the well test plan and the overall well test processes.

Contents

About the Author

Tarek Al-Arbi Omar Ganat is an accomplished associate professor with extensive experience in both the oil and gas industry and academia. He is a chartered engineer CEng (EI) member and was born in 1968. Dr. Ganat earned his Bachelor of Science degree in Petroleum Engineering from Tripoli University in Libya in 1991, followed by a Master of Science degree in Petroleum Engineering from ISE University in Spain/Madrid in 2003. He went on to obtain a Master of Science degree in Engineering Management from Tripoli University in 2005 and a PhD degree in Petroleum Engineering from the International Islamic University of Malaysia, Selangor, Malaysia, in 2016.

Dr. Ganat has over 28 years of work experience in the oil and gas industry, having worked with reputable organizations such as Jowef Oil services, Repsol Oil Company, Petro-Canada Oil Company, and PETRONAS Carigali Sdn in Malaysia. He has developed considerable expertise in well testing and reservoir engineering studies and application software development, with notable experience in the application of petroleum industry-related packages in multinational and multicultural joint ventures. In addition to his work in the industry, Dr. Ganat has made significant contributions to academia. He was one of the academic members of PETRONAS Teknologi Universiti in Malaysia for 5 years and a research engineer for the research institute of Hydrocarbon in Malaysia. He has also spent over 3 years in training centers, where he has developed professional skills and extensive experience.

Dr. Ganat has published over 83 papers in peer-reviewed journals and conferences and has authored three books that have been published by Springer. These include "Fundamentals of Reservoir Rock Properties" (2019), "Technical Guidance for Petroleum Exploration and Production Plans" (2020), and "Rock Properties Reservoir Engineering Practical View" (2022). Currently, he is a valued member of the petroleum and chemical engineering department at Sultan Qaboos University, where he continues to inspire and mentor the next generation of professionals in the field of petroleum engineering. His vast knowledge, extensive experience, and dedication to excellence make him a sought-after expert in his field.

Chapter 1
Introduction to Recognized Flow System During Well Testing

1.1 Fluid Flow in Porous Media

The concept of well testing arises with the interpretation of fluid flow in porous media. To have a porous medium in any rock type, it must contain spaces, known as voids or pores, free of solids, inserted in the solid matrix. The open holes are normally filled with some fluid such as gas, water, oil, or a mixture of all these fluids.

It must be leaky (permeable) to a mixture of fluids, i.e., fluids must be able to infiltrate across one side of a core sample and come out on the other side.

The reservoir engineer is concerned with the amounts of hydrocarbon content inside the rocks, the mobility of fluids inside the reservoir rocks, and reservoir rock properties. These properties depend on the type of reservoir rock and its distribution along permeable reservoir rock. Understanding the rock properties and the existing interface between the hydrocarbons and the reservoir rock is crucial to identifying and estimating the performance of the reservoir.

Typically, rock properties determined by conducting laboratory tests using real core samples from the reservoir in question. There are two important types of core analysis tests; Routine Core Analysis tests (RCA) to determine porosity, permeability, and saturation, and Special Core Analysis tests (SCAL), to determine capillary pressure relative permeability, wettability, surface and interfacial tension, electrical conductivity, and pore size distribution.

These properties represent the most important reservoir engineering parameters that help the reservoir engineers to describe the reservoir quality. Furthermore, these parameters are necessary for reservoir engineering studies as they strongly affect the quantity and distribution of fluids within the reservoir.

© The Author(s), under exclusive license to Springer Nature Switzerland AG 2023
T. A-A. O. Ganat, *Modern Pressure Transient Analysis of Petroleum Reservoirs*,
Petroleum Engineering, https://doi.org/10.1007/978-3-031-28889-0_1

1.2   Types of Flow Behaviour

1.2.1   Transient Flow

Pressure transient migrates outward from the well without encountering any boundaries. It is defined as a fluid dynamics condition. In this flow system, the pressure and the volume change over time because of changes in the system flow condition. Figure 1.1 shows the transient flow system around the wellbore the well boundary.

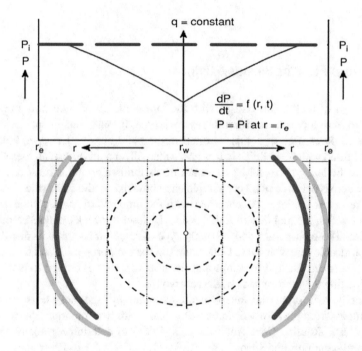

Fig. 1.1 Show the transient flow system diagram

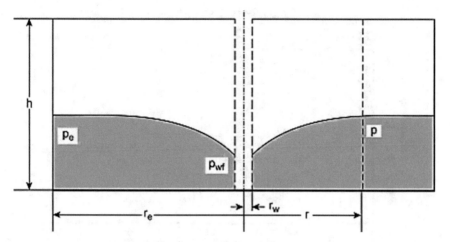

Fig. 1.2 A reservoir model illustrating a constant-pressure boundary

1.3 Steady-State Flow

Steady state flow describes the flow condition where the fluid properties (temperature, pressure, and velocity) at any location in the reservoir do not change with time. Also, the mass flow rate is constant in a steady-state flow system. This means that there is no accumulation of mass inside any component in the system. Figure 1.2, shows the reservoir model, and demonstrates the constant-pressure boundary (P_e) at distance (r_e) from the centre of the wellbore.

1.3.1 Pseudo-steady State Flow

This type of flow regime occurs in closed reservoirs when the pressure transient has moved to all the physical boundaries of the reservoir. The boundaries include also the surrounding producing wells. This flow test analysis is only valid for drawdown tests or injection data while the well is flowing. The pseudo-steady state analysis is not relevant for buildup or falloff tests. Figure 1.3 show the radius from the wellbore centreline to the affected reservoir region. This radius calls the radius of investigation (R_i). the radius of investigation increases as the well production time increases.

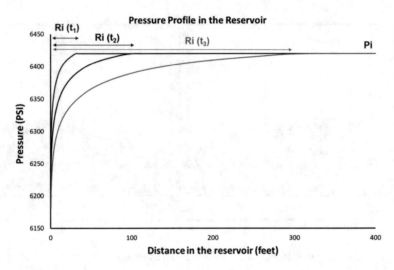

Fig. 1.3 Show the radius from the wellbore centreline to the affected reservoir region

1.3.2   Boundary-Dominated Flow

In the case of boundary-dominated flow, the pressure transient has touched all the boundaries. In this flow, the static pressure is decreases at the boundary, but not equal as the flow rate is not constant. Figure 1.4 show the boundary-dominated flow.

Fig. 1.4 Pressure depletion through the boundary-dominated flow

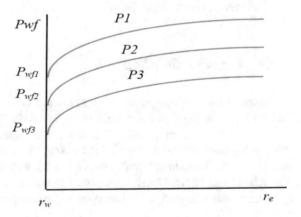

1.4   Development of the Diffusivity Equation

The diffusivity equation was generated by combining Darcy's law, conservation of mass law, and equations of state for the isothermal fluids flow of small and constant compressibility through a porous medium.

For radial flow on the way to the wellbore in a circular reservoir. The diffusivity equation is:

$$\frac{\partial^2 p}{\partial r^2} = \frac{1}{r}\frac{\partial p}{\partial r} = \frac{\phi\mu c}{0.0002637k}\frac{\partial p}{\partial t} \tag{1.1}$$

It's assumed that the diffusivity equation can be used for slightly compressible liquid, c, and independent of pressure. Where k is referred to the constant reservoir permeability (isotropic), viscosity, μ, P is independent of pressure, and porosity, ϕ, is constant. The hydraulic diffusivity is referred to the grouping term, $0.0002637\ k/\phi\mu c$.

1.5   Infinite-Acting Radial Flow

The best key solution of the diffusivity equation is the E_i-function solution, which represents infinite-acting radial flow. For a horizontal homogeneous reservoir, an infinite, having uniform initial pressure, with microscopic line-source well producing at the same flow rate, q, opening at time zero, the solution equation (Eq. 1.2) is as follows:

$$p(r, t) = p_i + \frac{70.6q\,B\mu}{kh}E_i\left(-\frac{9480\mu C_t r^2}{kt}\right) \tag{1.2}$$

where the E_i function is defined by Eq. 1.3:

$$E_i(-x) = \int_x^\infty \frac{e^{-y}}{y}\partial y \tag{1.3}$$

For fluid flow in porous media, the E_i function is zero for x > 10,

$$E_i(-x) \cong 0, \ x > 10 \tag{1.4}$$

For the argument, x, of the E_i function less than 0.01, the E_i function can be approximated by a logarithmic function (Eq. 1.5):

$$E_i(-x) = \ln(1.781x) \tag{1.5}$$

Equation 1.2 provides the pressure at time t and distance r from the centreline of the wellbore. Figure 1.5 illustrates a graph of the E_i function.

At any given time, t, or distance the Eq. 1.2 can be used to determine the pressure distribution in the reservoir area as a function of given distance or time, as shown in Figs. 1.3 and 1.6.

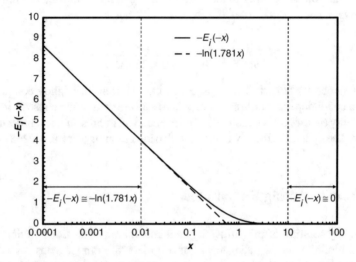

Fig. 1.5 Display the E_i-function solution for a line-source well

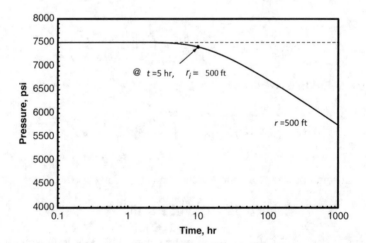

Fig. 1.6 Pore fluid pressure of shale in contact with sea water versus time at different distance

1.6 Concepts of Superposition

The principle of superposition identifies that the total pressure at any position in a reservoir area is the total of the pressure drops at that position affected by flow in each of the wells in the reservoir. The superposition principle indicates a pressure disturbance will spread across the reservoir even if the source of the disturbance changes.

1.6.1 Superposition in Time Versus Superposition in Space

In reservoir engineering, superposition is often carried out by two essential approaches in the analysis of pressure and flow rate data which are superposition in time and superposition in space.

When a well is producing at steady flow rate, q_1, for a certain time t_1 and then continue flow at different steady flow rate q_2 for time t_2, the required solution at time t_2 in this multiple-flow rate condition can be obtained by superposing the solution because of flow rate q_1 and the solution because of rate $(q_2 - q_1)$ at time $(t_2 - t_1)$. This approach called superposition in time. The same principle can be used to any number of varying flow rates.

When two wells (well-A and well-B) are producing from the same reservoir but at different positions, the solution at any distance in the reservoir is affected by the two producing wells. The required solution at location X can be obtained by superposing the solution at location X because of Well-A and the solution at location X because of Well-B. This approach called superposition in space. The same principle can be used to any number of well positions.

1.6.2 Superposing Pressures Versus Superposing Rates

When superposing the solutions concerning time, two methods can be used which are superposing pressures and superposing rates.

Superposing pressure responses presume that any solution's boundary condition is a constant flow rate. For instance, a well is flowing at a constant flow rate q_1 for a certain time t_1, and thereafter at a constant rate q_2 for time t_2, and lastly at a constant rate q_3 for time t_3. The total pressure drop at time t_3 (ΔP_t) is acquired by adding together the pressure responses due to the three constant rate condition:

$$\Delta P_t = \Delta P_1 + \Delta P_2 + \Delta P_3 \qquad (1.6)$$

where ΔP_1 caused by q_1 during the whole flow period (flow time $= t_3$), ΔP_2 caused by $(q_2 - q_1)$, starting at the time t_1 (flow time $= t_3 - t_1$), and ΔP_3 caused by at $(q_3 - q_2)$, starting at the time t_2 (flow time $= t_3 - t_2$).

Figures 1.7 and 1.8 shows a simplified multiple-rate example for superposing pressures. This used approach can be employed for any number of step flow rate changes (Liang 2015).

Also, superposing rates, presumes that any solution's boundary condition is constant pressure. For instance, a well is producing at constant pressure P_1 for time t_1, then at a constant pressure P_2 for time t_2, and finally at a constant pressure P_3 for time t_3. The total flow rate at the time (q_{t3}) is acquired by adding together the flow rate responses caused by the following three constant pressure conditions:

$$q_t = q_1 + q_2 + q_3 \tag{1.7}$$

where q_1 caused by $(P_i - P_1)$ during the whole flow period (flow time $= t_3$), q_2 caused by $(P_1 - P_2)$, starting at the time t1 (flow time $= t_3 - t_1$), and q3 caused by at $(P_2 - P_3)$, starting at the time t_2 (flow time $= t_3 - t_2$).

Fig. 1.7 Multiple-flow rate for superposing pressures

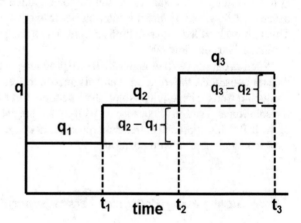

Fig. 1.8 Pressure profile for superposing pressures

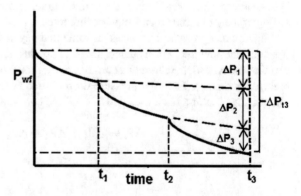

Fig. 1.9 Multiple-flow rate for superposing rates

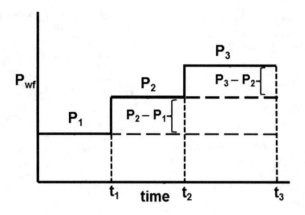

Fig. 1.10 Pressure profile for superposing rates

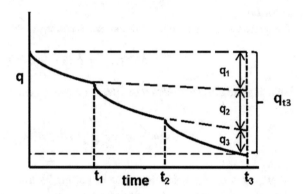

Figures 1.9 and 1.10 show an example of superposing rates. This multiple pressure approach can be used for any number of step pressure changes.

1.6.3 Example 1.1

In Fig. 1.11, the reservoir is producing with 3 oil wells. Well-A starts production at time $t_1 = 0$ with constant flow rate, q_1. Pressure drop was observed at observation well (well-C), $\Delta P_{3,1}(t)$. Well-B starts production later, at time t_2. If well-B, was the only production well, the pressure changes in well-C would be $\Delta P_{3,2}(t - t_2)$. If both wells produce, calculate the pressure change. By means of the dimensionless variables, write the equation for any number of wells.

Fig. 1.11 Show the locations of the three producing wells

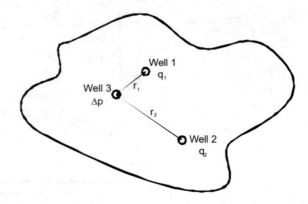

Solution

The pressure change can be calculated by adding together the pressure differences.

$$\Delta P(t, r) = \Delta P_{3,1} + \Delta P_{3,2}$$

By using the dimensionless variable, the equation can be written as follows:

$$\Delta P(t, r) = \frac{\mu}{2\pi hk} \sum_{j=1}^{n} q_j B_j p_D \big(t_D - t_{Dj}, r_{Dj} \big),$$

where r_{Dj} is the dimensionless time of putting induvial wells into operation and r_{Dj} is the dimensionless distance of wells from the location of the observation well.

1.6.4 Example 1.2

In an infinite acting reservoir, a well produces 40 m³/d (252.5 bbl/d) of oil for 5 days. After 5 days the flow rate reduced to $q = -25m^3$ (157.2 bbl/d). Calculate the bottom hole flowing pressure after 20 days. The given well data is as follows:
    The given well data as following:

Pi	33.24 MPa (4819.8 psi)
B_o	1.52 rbbl/bbl
μ_o	1.28 × 10⁻³ Pa(1.28 cP)
h	12 m (39.37 ft)
Flow rate	−40 m³/d = 0.463 × 10⁻³ m³/s (251.5 bbl/d)
t	5 days = 0.432 × 10⁶ s
K_o	0.16 × 10⁻¹² m² (160 mD)
ϕ	18%
C_t	3.86 × 10⁻⁹ Pa⁻¹ (2.662 × 10⁻⁵ 1/psi)

| r_w | 0.1 m (0.328 ft) |
| s | 23.97. |

Solution

t_1	0
t_2	5 days $= 0.432 \times 10^6$ s
t	20 days $= 1728 \times 10^6$ s
Δq_1	$q_1 = 40$ m³/d $= 0.463 \times 10^{-3}$ m³/s (251.5 bbl/d)
Δq_2	$q_2 - q_1 = 15$ m³/d $= 0.0174 \times 10^{-3}$ m³/s (94.33 bbl/d)
q_2	$- 25$ m³/d $= 0.000289$ m³/s (157.23 bbl/d).

The dimensionless time different according to the dimensionless time equation:

$$t_D = \frac{kt}{\emptyset \mu C_t r_w^2}$$

$t_D - t_{D1} = \frac{k(t-t_1)}{\emptyset \mu C_t r_w^2} = 17{,}991 \times 1728 \times 106 = 3109 \times 107$

$t_D - t_{D2} = 17{,}991 (1728 - 0.432) \times 10^6 = 2331 \times 10^7$
From dimensionless pressure equation,

$$P_{Dw}(t_D) = P_D(t_D, r_D = 1) + S$$

$$P_D(t_D - t_{D1}) = P_D(3109 \times 10^7) = 8.94$$

$$P_D(t_D - t_{D2}) = P_D(2332 \times 10^7) = 8.82$$

And the bottom hole flowing pressure after 20 days is:

$$P_{wf} = P_i + \frac{\mu B}{2\pi hk}[\Delta q_1 P_D(t_D - t_{D1}) + \Delta q_2 P_D(t_D - t_{D2}) + q_2 S]$$

$$P_{wf} = 33.24 \times 10^6 + \frac{1.28 \times 10^{-3} \times 1.52}{2\pi \times 12 \times 0.16 \times 10^{-12}}$$

$$\left[-0.463 \times 10^{-3} \times 8.94 + 0.174 \times 10^{-3} \times 8.82 - 0.289 \times 10^{-3} \times 24.34\right]$$

$$= 31.68 \text{ MPa}$$

Bottom hole flowing pressure in field unit:

$$P_{wf} = 4819.8 + 141.2 \times \frac{1.28 \times 1.52}{39.37 \times 160}$$

$$[-251.5 \times 8.94 + 94.55 \times 8.82 - 157.23 \times 23.97] = 4593.71 \text{ Psia}$$

1.7   Radius of Investigation

Radius of investigation reveals how far into the reservoir zone the pressure transient effects have travelled. It is one of the best practical concepts in well test interpretation.

A pressure transient is established when an interruption such as a change in flow rate occur at a wellbore. As time continues, the pressure transient progresses more and more into the reservoir. Typically, when a pressure disturbance is started at the wellbore, it will have instant effect in the area nearby the wellbore and insignificant effect at all locations in the reservoir which are out for the well drainage area. The maximum distance at which the pressure transient effect in the reservoir is known as the radius of investigation, r_{inv}.

Reserve estimation was a crucial issue as it is being applied to justify the development of any project. Generally, there are many different drainage area estimation approaches have been developed and applied but there is no a specific method was strongly recommended to be applied in oi and gas industry for reserve calculation (Du 2007; Fikri and Kuchuk 2009). The conventional radius of investigation (r_{inv}) is expressed in the Eq. 1.8 below:

$$r_{inv} = \sqrt{\frac{kt}{9480\phi\mu c_t}} \tag{1.8}$$

where:

t time, hours
k permeability, mD
c_t total compressibility, psi^{-1}

Figure 1.12, visualizes the pressure transient progresses for radial flow and for a vertical well with an infinite conductivity fracture with increasing times. The well produces at a constant flow rate from a reservoir initially at a uniform pressure. Figure 1.13 displayed the radius of investigation for a horizontal well with multistage hydraulic fractures reservoir at various times.

1.7.1   Example 1.3

Find the radius of investigation when the permeability is 12 mD, porosity is 15%, viscosity is 1cP, time is 7 h and total compressibility is 6.5×10^{-6} Psi^{-1}.

Solution
The radius of investigation equation is:

$$r_{inv} = \sqrt{\frac{kt}{9480\phi\mu c_t}}$$

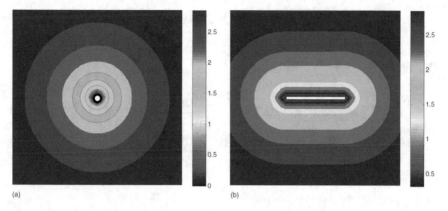

Fig. 1.12 Radius of investigation at various times in days **a** for radial flow and **b** for a vertical well with an infinite conductivity fracture

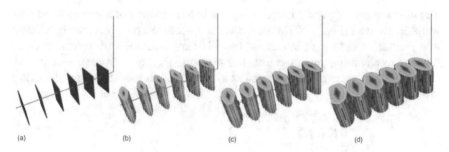

Fig. 1.13 Radius of investigation for a horizontal well with multistage hydraulic fractures reservoir at various times **a** at the initial day **b** after 3 months **c** after 6 months, and **d** after one year

$$r_{inv} = \sqrt{\frac{12 \times 7}{948 \times 0.15 \times 1.0 \times 3.5 \times 10^{-6}}} = 411 \text{ ft}$$

1.7.2   Example 1.4

If pressure disturbance reaches a distance of 275 ft at the reservoir rock and fluid data:

K 10 mD,
ϕ 16%,
μ 1.04 cP, and
C_t 6.5 × 10^{-6} Psi^{-1}

Calculate the time required to reach this distance.

Solution
The time is:

$$t = \frac{9480\mu C_t r_{inv}^2}{k}$$

$$t = \frac{948 \times 0.16 \times 1.04 \times 6.5 \times 10^{-6}(275)^2}{10} = 7.75\,\text{h}$$

1.7.3 Example 1.5

The pressure results for a 7-h injectivity test in a reservoir that has been filled with water are shown in Fig. 1.14. Before the test began, the reservoir had been flooded for two years at a steady injection rate of 100 STB/day. Once all wells were shut down for four weeks to maintain the reservoir pressure, the injectivity test was started. Utilizing the given reservoir rock and fluid parameters below, determine the radius of investigation:

c_t	6.67×10^{-6} psi^{-1}
B	1.0 bbl/STB,
μ	1.0 cP
Sw	62.4 lb/ft3,
ϕ	0.15,
q_{inj}	100 STB/day
h	16 ft,
r_w	0.25 ft,
p_i	194 psig
ΔSw	0.4,
Depth	1002 ft,
Total test time	7 hours

Solution
First, determine the permeability using the following equation:

$$K = \frac{162.6\, q_{inj} B\mu}{mh}$$

From Fig. 1.1, the slope "m" is 80 psig/cycle.

$$K = \frac{162.6 \times 100 \times 1.0 \times 1.0}{80 \times 16} = 12.7\,\text{mD}$$

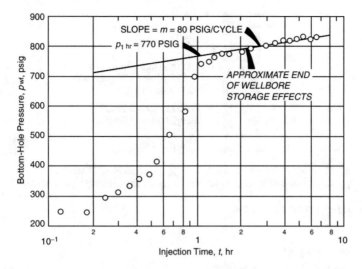

Fig. 1.14 Shows pressure recorded data for a 7-h injectivity test in a water-flooded reservoir

The radius of investigation equation is:

$$r_{inv} = \sqrt{\frac{kt}{948 \varnothing \mu C_t}}$$

$$r_{inv} = \sqrt{\frac{12.7 \times 7}{948 \times 1.0 \times 0.15 \times 6.67 \times 10^{-6}}} = 306\,\text{ft}$$

1.8 Altered Zone and Skin Factor

During drilling operations, most drilled wells have decreased their permeability (due to formation damage) in the vicinity of the wellbore caused by invaded of drilling or completion fluids (Fig. 1.15). Many other wells are stimulated by acidization or hydraulic fracturing. Any changes in the reservoir permeability in the area around the wellbore radius will seriously invalidate the basic radial flow forms. Therefore, the basic radial flow equations must be modified by including a correction term recognized as the skin factor which represents the permeability alteration phenomena.

Normally, the permeability damage can be caused by:

- Mud solids invasion
- Clay swelling
- Chemical and mechanical treatment of Sand consolidation
- Particle tilting

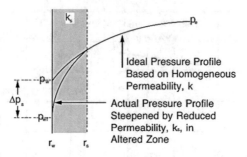

k_s . . . Altered Zone Permeability   r_s . . . Extent of Alteration
p_wf . . Actual Bottom-hole Pressure   Δp . . . Incremental Pressure Drop

Fig. 1.15 Near wellbore damage zone

- Formation compaction
- Chemical precipitation or scaling
- Movement of fine grains near the wellbore.

Injection wells are also causing formation damage combined with:

- Plugging as a result of particulate matter in the injection fluid
- Clay alteration on contact with injected fluid
- Injected water incompatibility with the formation of water

Certainly, always there is potential to improve the permeability in the area around the wellbore by different stimulation methods such as acidizing and fracturing.

In developing the infinite-acting radial flow equation (Eq. 1.2), it is presumed that the well is vertical, has been completed as an open hole well, and is not damaged or stimulated. In formulating an infinite-acting radial flow equation (Eq. 1.2), it is assumed that the well is vertical, completed as an open hole well, and is neither damaged or stimulated. There are several factors that explain formation damage or stimulation, cased hole and perforated or gravel-pack completion, partial penetration, and the skin component, or departure from vertical (Van Everdingen 1953). Each of these scenarios has the primary effect of changing the pressure at the wellbore from what Eq. 1.2 anticipated.

1.9   Modeling Skin of Altered Permeability Area

The damaged or simulated region around the wellbore is often assumed to be created by an annulus of changed permeability around the well, but the permeability of the reservoir distant from the wellbore stays constant (Van Everdingen 1953; Hawkins 1956). This implies that if there is damage, the permeability in the changed zone will be less than the reservoir permeability, but if the permeability in the altered zone is more than the reservoir permeability, the well is stimulated. Figure 1.16 depicts the damaged region surrounding the wellbore, as well as a simplified well model layout. The damaged or stimulated annular area has an inside radius r_w, outside radius ra, and homogeneous permeability k_a.

Figure 1.17a illustrates the pressure gradient around the wellbore where there is no change in the original reservoir permeability. In this case, the pressure drop through the altered zone annulus is expressed as:

$$\Delta p_u = \frac{141.2q\,B\mu}{kh} \ln \frac{r_a}{r_w} \tag{1.9}$$

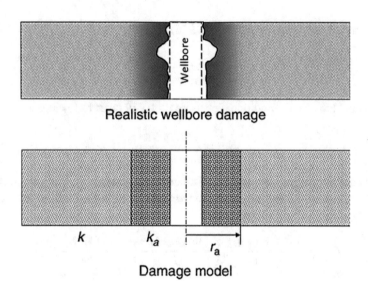

Fig. 1.16 Realistic and idealized models for near-wellbore damage

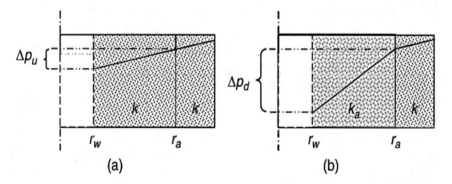

Fig. 1.17 Determining the further pressure drop through the altered zone

In the case of the altered zone present, as shown in Fig. 1.17b, the pressure drop is expressed as:

$$\Delta p_d = \frac{141.2q B \mu}{k_a h} \ln \frac{r_a}{r_w} \tag{1.10}$$

The additional pressure drop generated by damage or stimulation is expressed by:

$$\Delta p_s = \Delta p_d - \Delta p_u = \frac{141.2q B \mu}{kh} \left(\frac{k}{k_a} \right) \ln \frac{r_a}{r_w} \tag{1.11}$$

Skin factor as defined by Hawkins (1956) is:

$$S = \left(\frac{k}{k_a} - 1 \right) \ln \frac{r_a}{r_w} \tag{1.12}$$

The final look for the additional pressure drop will be as following:

$$\Delta p_s = \frac{141.2q B \mu}{kh} S \tag{1.13}$$

The bottom hole pressure can be calculated by assessing Eq. 1.2 at the wellbore and adding the additional pressure drop created by skin factor from Eq. 1.13 as follows:

$$p(t) = p_i + \frac{70.6q B \mu}{kh} E_i \left(-\frac{948 \emptyset \mu C_t r^2}{kt} \right) - \frac{141.2q B \mu}{kh} S \tag{1.14}$$

1.9.1   Example 1.6

Calculate the formation damage permeability for the following given well data:

- Initial reservoir permeability = 120 mD
- Damaged well bore radius = 4 ft
- Wellbore radius = 0.3 ft
- Skin factor = 26

Solution
Sink factor equation defined as:

$$S = \left(\frac{k}{k_a} - 1 \right) \ln \frac{r_a}{r_w}$$

$$26 = \left(\frac{120}{k_a} - 1 \right) \ln \frac{4}{0.3}$$

Which on solving for k_a gives 10.87mD.

1.9.2 Example 1.7

Using the same well data given in example 1.6, what will be the value of the skin factor if $k_a = 108$ mD?

Solution
The Hawkins equation is:

$$S = \left(\frac{k}{k_a} - 1 \right) \ln \frac{r_a}{r_w}$$

$$S = \left(\frac{120}{108} - 1 \right) \ln \frac{4}{0.3} = 0.288$$

1.10 Wellbore Storage (WBS)

The compressibility of fluid in the wellbore causes wellbore storage, which prolong the period of pressure response.

The evaluation of the wellbore pressure response during drawdown (DD) and buildup (BU) tests is known as the well test analysis. In most cases, the wellhead rather than the sand face controls the flow rate. Since the fluid was compressed in the wellbore before the well opened, the steady flow rate at the wellhead while the well is controlled from the surface does not imply that the reservoir is producing at its full rate. The wellbore storage is the cause for this result. In a drawdown test, the well is allowed to flow after being shut in, which causes a decrease in wellbore pressure. The two types of wellbore storage are as follows:

- Wellbore storage effect caused by fluid expansion at the wellbore.
- The increasing fluid level in the casing-tubing annulus causes a wellbore storage effect.

The wellbore storage effect during the following two tests:

Drawdown test

- The early flow rate is caused by wellbore unloading when the well is opened at the surface (Fig. 1.18).

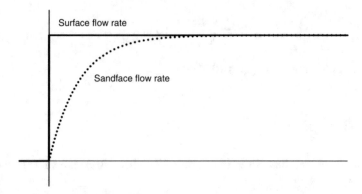

Fig. 1.18 Wellbore storage, DD test, steady surface flow rate. Sand face flow rate increases gradually, reaching surface flow rate

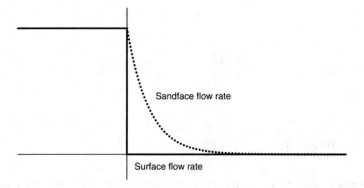

Fig. 1.19 Wellbore storage, BU test. Sand face flow rate reduces gradually to zero

- As wellbore unloading slowly drops, the flow from the reservoir to the wellbore rises gradually to reach the required flow rate at the wellhead (q_{wh}).

Buildup test

- When the well is closed at the surface, flow from the reservoir does not stop directly (Fig. 1.19).
- After closing the well, the flow of fluid from the reservoir to the wellbore continues for a certain time due to the compressibility of the fluid.
- The flow rate at the wellhead decreases gradually to zero during the shut-in period.

1.11   Fluid-Filled Wellbore (WBS)

A well filled with a single-phase fluid is shown in Fig. 1.20. Because the pressure at the wellhead is always lower than the bubble point pressure, oil wells closed from the surface valve often do not indicate a liquid-filled wellbore. A fluid-filled wellbore can develop in a water injection well after the pump has been turned off to conduct a falloff test. The relation between surface and sand face flow rates in a wellbore filled with a single-phase fluid is demonstrated by the material balance equation below:

$$24C_{wb}V_{wb}\frac{dp_w}{dt} = -\big(q(t) - q_{sf}(t)\big)B_{wb} \tag{1.15}$$

where c_{wb} is the wellbore fluid compressibility in psi^{-1}, V_{wb} denoted the wellbore volume in bbl, q_{sf} is the flow rate at a sand face in stb/d, p_w is the wellbore pressure in psi, and, B_{wb} is the formation volume factor of the fluid in bbl/stb.

The wellbore storage coefficient for fluid expansion, $C_{FE,}$ is defined as:

$$C_{FE} = -c_{wb}V_{wb} \tag{1.16}$$

So, Eq. 1.15 can written as

$$24C_{FE}\frac{dp_w}{dt} = -\big(q(t) - q_{sf}(t)\big)B_{wb} \tag{1.17}$$

Equation 1.17 can be used for variable flow rate production, DD, and BU tests. Where before the start of DD test, both the surface flow rate and sand face flow rate are equal to zero. Once the surface choke valve is opened, the surface flow rate increases immediately to a constant flow rate value. While the pressure rate will change instantly when the choke valve is opened, when the sand face flow rate is almost zero, is then given by:

Fig. 1.20 Wellbore storage, fluid-filled from reservoir into wellbore

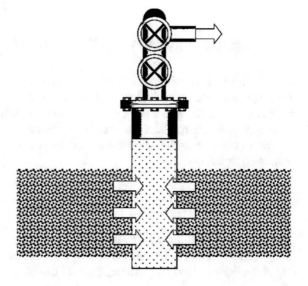

$$24C_{FE}\frac{dp_w}{dt} \cong -(q-0)B_{wb}$$

$$24C_{FE}\frac{dp_w}{dt} = -qB_{wb} \tag{1.18}$$

The below bottom pressure, p_{wf}, in this wellbore storage dominated period, is then given by:

$$p_i - p_{wf} \cong \frac{qB_{wb}}{24C_{FE}}t \tag{1.19}$$

where p_{wf} is the flowing bottomhole pressure at the time shut-in the well.

Also, before shutting in for a BU test, both sand face and surface flow rates have the same rate. Immediately after closing in the well before the sand face flow rate has started to drop, the rate of pressure change in the wellbore is given by:

$$24C_{FE}\frac{dp_w}{dt} \cong -\left(0-q_{sf}\right)B_{wb}$$

$$24C_{FE}\frac{dp_w}{dt} = q_{sf}B_{wb} \tag{1.20}$$

And the shut-in bottom pressure, p_{ws}, is expressed by:

$$p_{ws} - p_{wf} \cong \frac{qB_{wb}}{24C_{FE}}\Delta t \tag{1.21}$$

1.12 Falling Liquid Level in the Wellbore

As seen in Fig. 1.21, oil wells often have a filled liquid level rather than a wellbore filled with a single-phase fluid. This phenomenon occurs in oil wells having sucker rod pumps, when the well is pumped off, and in water-injection wells, where the well goes on vacuum shortly after the pump is turned off.

The wellbore material balance equation for a well with a filling liquid level is expressed as follows:

$$24\frac{144}{5.615}\frac{A_{wb}}{\rho_{wb}}\frac{d(p_w - p_t)}{dt} = -\left(q - q_{sf}\right)B_{wb} \tag{1.22}$$

where ρ_{wb} denotes liquid density in lbm/ft^3, p_t denotes the pressure at the top of the liquid column, and A_{wb} denotes the area of the wellbore in ft^2. If the pressure at the top of the liquid column is constant, Eq. 1.21 can be expressed as:

Fig. 1.21 Wellbore storage, rising liquid level

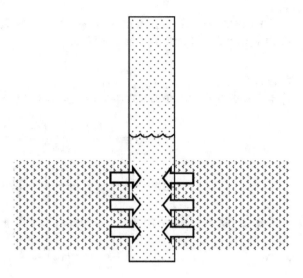

$$24\frac{144}{5.615}\frac{A_{wb}}{\rho_{wb}}\frac{dp_w}{dt} = -\left(q - q_{sf}\right)B_{wb} \qquad (1.23)$$

If the liquid level changes, the WBS coefficient equation for filled liquid can be written as:

$$C_{FL} = \frac{144}{5.615}\frac{A_{wb}}{\rho_{wb}} = -25.65\frac{A_{wb}}{\rho_{wb}} \qquad (1.24)$$

In this case, Eq. 1.23 is simplified and written as:

$$24C_{FL}\frac{dp_w}{dt} = -\left(q - q_{sf}\right)B_{wb} \qquad (1.25)$$

1.12.1 Example 1.8

Below are the well data for an oil well that is planned to perform a drawdown test.
    The volume of fluid in the wellbore $= 180$ bbl,

C_o 10×10^{-6} psi^{-1},
ρ_o 45 lb/ft^3.
Tubing ID   2-in and OD $= 7.675$-in.

    Calculate:

1. Annulus cross-sectional area, A_a.
2. The wellbore storage factor caused by fluid expansion.

3. The wellbore storage factor is produced by the falling fluid level.
4. The overall wellbore storage coefficient.

Solution

Step-1. Calculate the cross-sectional area of the annulus A_a:

$$A_a = \frac{\pi\left[(ID_C)^2 - (ID_T)^2\right]}{4(144)}$$

$$A_a = \frac{\pi\left[(7.575)^2 - (2)^2\right]}{4(144)} = 0.2995 \text{ ft}^2$$

Step-2. Estimate the WBS factor produced by fluid expansion:

$$C_{FE} = -c_{wb}V_{wb}$$

$$C_{FE} = -(180)(10 \times 10^{-6}) = 0.0018 \text{ bbl/psi}$$

Step-3. Calculate the WBS factor produced by the falling fluid level:

$$C_{Fl} = \frac{144}{5.615}\frac{A_{wb}}{\rho_{wb}} =$$

$$C_{Fl} = \frac{(144)(0.2995)}{(5.615)(45)} = 0.1707 \text{ bbl/psi}$$

Step-4. Determine the total wellbore storage coefficient:

$$C = C_{FE} + C_{Fl}$$

$$C = 0.0018 + 0.1707 = 0.1725 \text{ bbl/psi}$$

Note: The findings demonstrate that the impact of fluid expansion CFE may be disregarded in crude oil systems.

1.13 Summary

Fluid flow in a porous media is influenced by a variety of factors, and its primary function is to consume energy and produce fluid via the wellbore. In flow dynamics through porous media, the connection between energy and flow rate becomes the most critical concern. The fluid flow during the well testing procedure may interpret many fluid flow systems. This chapter covers the principles of fluid flow theory in porous media for a variety of systems. The diffusivity equation is presented for transient pressure behaviour with appropriate pressure solutions. Superposition, boundary-dominated flow, wellbore storage (WBS), and skin factor have also been comprehensively addressed. This chapter provides solutions to the exercises offered in different sections in order to emphasize the procedures mentioned in each section, assisting well test engineers in better understanding the fluid flow mechanism and the overall well test process.

References

Du, K. (2007). Use of Advanced Pressure Transient Analysis (PTA) Techniques to Improving Drainage Area Calculations and Reservoir Characterization: Field Case Studies. Society of Petroleum Engineers. SPE-109053-MS.

Fikri J. Kuchuk. (2009). Radius of Investigation for Reserve Estimation from Pressure Transient Well Tests. SPE-120515-MS. https://doi.org/10.2118/120515-MS.

Hawkins, M.F. Jr. 1956. A Note on the Skin Effect. J. Pet Tech 8 (12): 65–66. SPE-732-G. https://doi.org/10.2118/732-G.

Liang, Y. (2015). A New Method for Production Data Analysis Using Superposition-Rate (Unpublished master's thesis). University of Calgary, Calgary, AB. https://doi.org/10.11575/PRISM/24870.

Van Everdingen, A.F. 1953. The Skin Effect and Its Influence on the Productive Capacity of a Well. J. Pet Tech 5 (6): 171–176. SPE-203-G. https://doi.org/10.2118/203-G.

Chapter 2
Well Test Concepts

2.1 Flow Regimes and the Diagnostic Plot

One of the most crucial steps in identifying the reservoir model is describing the form of flow regime that is present during a well test. The use of straight-line analysis depends on the presence of a certain flow regime. To do a successful straight-line analysis, the beginning and end of each flow regime must be identified. On the following plots, every single flow regime behaviour can be seen:

- Diagnostic Log–log plot.
- Flow regime specific specialized diagnostic plot.
- Flow regime specific straight-line plot.

The main log–log diagnostic plot shows the pressure change and log–log derivative of pressure vs time on a log–log scale. As seen in Fig. 2.1 and reported in Table 2.1, the log–log derivative will have a distinct slope for each flow regime.

The horizontal line for data that displays the flow regime allows for easy graphic diagnosis of the flow regime, which is one of the key advantages of the flow regime particular specialized diagnostic plot. Additionally, make it possible to quickly evaluate each flow regime's slope from the horizontal line of the diagnostic plot for that particular flow regime.

The flow regime specific straight-line plot should be used to validate the appropriate flow regime identification established by both the log–log diagnostic plot and the flow regime specific diagnostic plot. The start and end of the flow time for every flow regime should then be determined using the log–log diagnostic plot or the diagnostic plot related to that flow regime, and then transferred to the straight-line plot for that flow regime. Typically, the data on the straight-line plot that is separate from the flow regime must display a straight line. If the presented data doesn't quite consist of a single straight line, double-check the pressure data and the derivative computation.

The log–log diagnostic plot is common to every flow regimes, however the specialized diagnostic and straight-line plots are distinct to each flow regime. The log–log

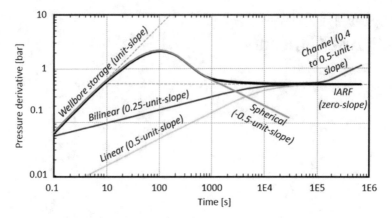

Fig. 2.1 Log–log plot showing characteristic responses of various flow regimes

Table 2.1 Different flow regime diagnostic characteristics

Flow regime	Slope of logarithmic derivative on log–log plot	Flow-regime specific diagnostic plot	Flow-regime specific straight-line analysis plot
Linear	1/2	$\frac{\partial p}{\partial t^{1/2}}$ versus Δt	p versus $\Delta t^{1/2}$
Bilinear	1/4	$\frac{\partial p}{\partial t^{1/4}}$ versus Δt	p versus $\Delta t^{1/4}$
Volumetric	1	$\frac{\partial p}{\partial t}$ versus Δt	p versus Δt
Radial	0	$\frac{\partial p}{\partial \ln t}$ versus Δt	p versus $\log \Delta t$
Spherical	$-1/2$	$\frac{\partial p}{\partial t^{-1/2}}$ versus Δt	p versus $\Delta t^{-1/2}$

diagnostic plot could be used to assess and estimate the beginning and end of each flow regime present during a test.

Currently, the log–log diagnostic plot is the best accessible approach for characterizing infinite-acting radial flow (IARF). Besides, the log–log diagnostic plot could be used to characterize other flow systems such as linear flow, bilinear flow, spherical flow, pseudo steady sate flow, and wellbore storage.

2.2   Power-Law Function

The pressure response for all flow regimes can be expressed in the form of a power-law function along with a constant:

$$\Delta p = m_n t^n + b_n, \tag{2.1}$$

where m_n is the slope and bn is the intercept of Δp versus t_n plot, n is a characteristic exponent specific for every single flow regime. Typically, the b_n is associated with a skin factor for all types of flow regimes. The log–log graph will display the pressure change as a straight line with a slope of n when the intercept bn equals zero. The pressure change will resemble a straight line with a slope of n if b_n is not equal to zero and n is greater than zero. Equation 2.1 does not characterize the pressure change form during the radial flow, the log–log derivative through radial flow is a power-law function with $n = 0$.

2.3 Log–Log Derivative of a Power-Law Function

If logarithm was applied on both sides of the power-law function Eq. 2.1 at $b_n = 0$, the equation will be written as:

$$\log(\Delta p) = \log(m_n t^n) = n\log(t) + \log(m_n) \tag{2.2}$$

Therefore, when b_n is equal to zero, a log–log plot will show a straight line with slope n and intercept log (m_n). If logarithm derivative was applied on both sides of the power-law function Eq. 2.1 at $b_n = 0$, the equation will be written as:

$$t\frac{\partial p}{\partial t} = t\left(nm_n t^{n-1}\right) = nm_n t^n \tag{2.3}$$

By applying the logarithm on both sides of Eq. 2.3, the Eq. 2.3 can be written as:

$$\log\left(t\frac{\partial p}{\partial t}\right) = \log\left(nm_n t^n\right) = n\log(t) + \log(nm_n) \tag{2.4}$$

The log–log plot of $(t\,\partial p/\partial t)$ versus (t) will give an intercept log (nm_n) and a slope n. The slope n is utilized to identify flow regimes characterized by Eq. 2.1.

Usually, in order to describe a flow regime, the log–log derivative must be used, and the pressure change curve serves just as a consistency check. The pressure changes and pressure derivative curves are parallel, and both curves have a slope of n, according to the log–log diagnostic plot, if b_n is equal to zero and n is greater than zero. The derivative curve will be $1/n$ units higher than the pressure change curve. The pressure change curve will reach a line of slope n and be concave upward if both b_n and n are greater than zero. This line will be shifted upward from the derivative curve by a ratio of $1/n$ (Spivey 2013).

Equation 2.1 cannot apply to the pressure recorded during radial flow. As an Alternative, the pressure response has the following form:

$$\Delta p = m \, \log t + b \tag{2.5}$$

As a result, the power-law function below provides a constant that represents the log–log derivative for radial flow $n = 0$:

$$t\frac{\partial p}{\partial t} = t\left(\frac{2.303\,m}{t}\right) = 2.303\,m \tag{2.6}$$

2.4   Flow-Regime—Specific Diagnostic Plots

Every type of flow regime can be defined with Eq. 2.1. By using the derivative of Eq. 2.1 concerning t^n, the flow-regime-specific diagnostic plot can be defined as follows:

$$\frac{\partial p}{\partial t^n} = m_n \tag{2.7}$$

Radial flow can be easily identified on the diagnostic plot because the derivative curve is horizontal at the radial flow period.

Therefore, the flow-regime-specific diagnostic plot, defined as a log–log plot of dp/dtn versus t, will show a horizontal line with a constant intercept value. The flow-regime-specific diagnostic plots consist of all types of diagnostic plots as shown in Fig. 2.1. It is a graph that indicates the performance of the derivative as a function of the type of flow behaviour. The graph can be superposed to the data and shifted to recognize the type of flow that appears during the test period. Ehlig-Economides et al. (1994) provided many examples of the use of this method. The flow-regime-specific derivative can be determined from the log–log derivative by using the following equation:

$$\frac{\partial p}{\partial t^n} = \frac{1}{nt^{n-1}}\frac{\partial p}{\partial t} = \frac{1}{nt^n}\left(t\frac{\partial p}{\partial t}\right) \tag{2.8}$$

2.5   Flow-Regime-Specific Straight-Line Analysis Plots

The flow regime specified by Eq. 2.1 is acquired by plotting the wellbore pressure versus t^n. The time plotting function must be modified to take superposition into account during a buildup or multi-rate test. The plot's slope and intercept are used to calculate the reservoir's characteristics. The flow-regime-specific straight-line analysis plot for radial flow is just a semi-log plot.

2.6   Flow Regimes

Typically, the fluid in the pore's media flows in several ways at different times and it is depending also on the reservoir extent and shape (Fig. 2.2). This section will discuss the fundamental flow regimes that are classified based on the time region they occur, and the type of wellbore used to penetrate the reservoir, (vertical or horizontal).

Figure 2.3 show the standard derivative pressure–time plot with the different time regions observed (early time (E.T.), steady state (S.S.), and pseudo-steady state (P.S.S.)):

Flow regimes that appear within each of the flow regime types are scheduled in Table 2.2 based on the type of wellbore configuration.

2.6.1   Radial Flow

In the redial flow regime, flow is in the horizontal radial path (Fig. 2.4). This flow appears in the time region before the pressure transient has touched the boundaries of the reservoir (infinite-acting time).

The pressure response in the radial flow regime is a linear function of the logarithm of time, having the following form:

$$\Delta p = m \log(t) + b \tag{2.9}$$

Fig. 2.2   Derivative log–log diagnostic plot illustrating different time regions during the test

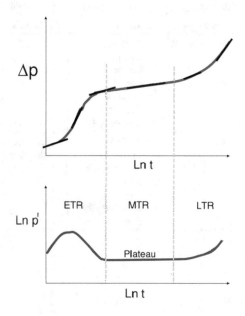

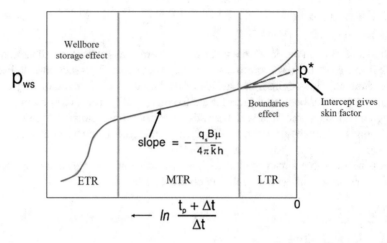

Fig. 2.3 Derivative and pressure–time plots

When plotting pressure against time on a semi-log graph, the pressure response data in the radial flow regime will follow a straight line with a slope of m. The geometric mean permeability in the plane where radial flow is taking place is inversely proportional to the slope m. In the Infinite-Acting Redial Flow (IARF) time, Fig. 2.5 depicts the log–log derivative as a horizontal line. The slope m can be determined

Table 2.2 show flow regime based on type of well

Wellbore	Early time	Middle time	Transition	Late time
Vertical wells	• Wellbore storage • Linear fracture flow • Bilinear fracture flow • Spherical flow	• Radial flow	• Single no flow boundary • Linear channel flow	• Pseudo-steady state flow • Steady state flow
Horizontal wells	• Wellbore storage • Vertical radial flow • Linear horizontal flow • Elliptical flow	• Horizontal radial flow	• Linear channel flow	• Pseudo-steady state flow • Steady state flow

(continued)

Table 2.2 (continued)

Wellbore	Early time	Middle time	Transition	Late time
Multi-fractured horizontal wells (MFHWs)	• Wellbore storage • Vertical radial flow within the fractures • Linear flow within the fractures • Bilinear flow	• Early linear flow (toward fractures) • Early radial flow (around each fracture prior to interference between fracs) • Compound linear flow • Late radial flow (around MFHW and fracture network)		• Pseudo-steady state flow

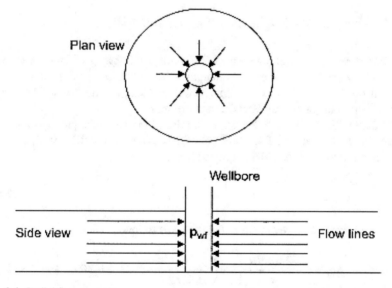

Fig. 2.4 Redial flow geometry

from a semi-log plot of pressure versus time, or the horizontal line on the log–log diagnostic plot. The geometric mean permeability can be calculated from slope m.

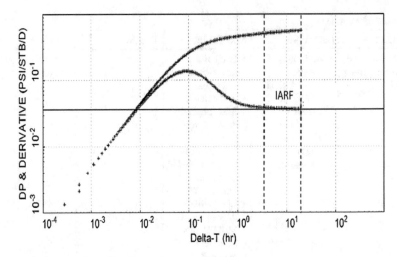

Fig. 2.5 Diagnostic plot showing IARF

2.6.2   Vertical Well, Infinite-Acting Redial Flow

The fluid flows to the wellbore equally from all paths throughout the IARF flow time, and the pressure drop extends radially. The wellbore is perpendicular to the reservoir boundaries, the reservoir thickness is uniform, and the top and bottom of the reservoir boundaries are parallel and delineated.

The initial IARF regime is known as infinite-acting when the pressure response reached the first boundary. The IARF pressure response in a vertical well, Fig. 2.6, is presented in the dimensionless equation by:

$$p_D = \frac{1}{2}\ln(1.781t_D) + S \tag{2.10}$$

The equation can be written in oilfield units as follows:

$$p_i - p_{wf} = \frac{162.6\,qB\mu}{kh}\left[\log\left(\frac{kt}{\emptyset\mu C_t r_w^2}\right) - 3.23 + 0.869\ S\right] \tag{2.11}$$

The logarithmic derivative equation is expressed as follows:

$$t_D\frac{\partial P_D}{\partial t_D} = \frac{1}{2} \tag{2.12}$$

The logarithmic derivative equation in dimensionless form, and oilfield units is expressed as:

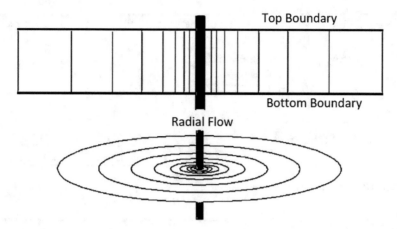

Fig. 2.6 Infinite-acting radial flow regime

$$t\frac{\partial p}{dt} = \frac{70.6\,q\,B\mu}{kh} \tag{2.13}$$

2.6.3   *Vertical Well Near Single No-Flow Boundary, Hemiradial Flow*

Typically, the no-flow boundary is a border that does not permit flow through it. The nature of such boundaries exists in reservoirs with sealing faults, or it is occurring in the middle distance between two producing wells or injecting wells that are producing/injecting at the same flow rate. Pseudo-steady state flow indicates that all no-flow boundaries have been touched.

Figure 2.7 illustrates how to simulate a case where a well is near to a sealing fault by eliminating the fault and placing an image well with the same flow rate as the producing or injection well.

The equation of dimensionless pressure response through hemiradial flow (HRF) with a single no-flow boundary (as seen in Fig. 2.8), for a vertical well, can be expressed as:

$$p_D = \ln(1.781\,t_D) + S + S_{HRF} \tag{2.14}$$

The equation can be written in oilfield units as;

$$p_i - p_{wf} = \frac{325.5\,q\,B\mu}{kh}\left[\log\left(\frac{kt}{\emptyset\mu C_t r_w^2}\right) - 3.23\right] + \frac{141.2\,q\,B\mu}{kh}\,(S + S_{HRF}) \tag{2.15}$$

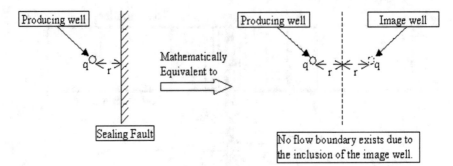

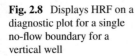

Fig. 2.7 No flow boundary examples

Fig. 2.8 Displays HRF on a diagnostic plot for a single no-flow boundary for a vertical well

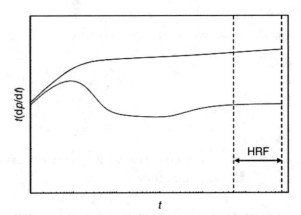

where S_{HRF} is a negative geometric skin factor expressed as:

$$S_{HRF} = \ln \frac{2L}{r_w} \tag{2.16}$$

where L is the distance to the single boundary

2.6.4 Vertical Well Between Intersecting Sealing Faults

Naturally, the intersecting sealing faults are the boundaries that do not allow flow to cross them. These types of boundaries are identified when the vertical well is placed between two or more sealing faults. Figure 2.9 exhibits the dimensionless pressure response for a vertical well located between intersecting sealing faults (fractional radial flow (FRF)). Equation 2.17 expressed the dimensionless pressure response as:

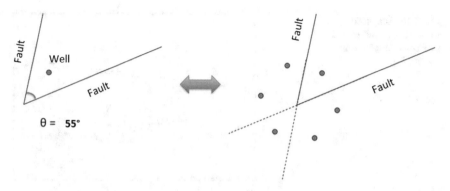

Fig. 2.9 Intersect faults examples

$$p_D = \frac{\theta}{180} \ln(1.781 t_D) + S + S_{FRF} \tag{2.17}$$

where the geometric skin factor is S_{FRF} and θ is the angle between two sealing faults (intersect the angle).

The dimensionless pressure response equation can be written in oilfield units as:

$$p_i - p_{wf} = \frac{0.9033\, q B \mu}{kh\theta} \left[\log\left(\frac{kt}{\emptyset \mu C_t r_w^2} \right) - 3.23 \right]$$
$$+ \frac{141.2\, q B \mu}{kh} (S + S_{HRF}) \tag{2.18}$$

The geometric skin factor can be calculated using the following equation if $\theta = 360°$ is an integer multiple of q:

$$S_{FRF} = \sum_{i=1}^{n-1} \ln \frac{L_i}{r_w} \tag{2.19}$$

where L_i denotes the distance to the ith image well and $n = 360/q$. If the well is located halfway from the two sealing faults, then n could be odd or even. For instance, if the well is closer to one of the intersecting sealing faults than to the other fault, n has to be even.

If the well is halfway from the two intersecting sealing faults, Eq. 2.19 can be written as:

$$S_{FRF} = - \ln \frac{L}{r_w} - \frac{1}{2} \sum_{i=1}^{n-1} \ln\{ [\text{Cos}(i\theta) - 1]^2 + \text{Sin}^2 i\theta \} = - \ln \frac{nL}{r_w} \tag{2.20}$$

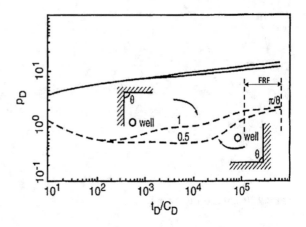

Fig. 2.10 Diagnostic plot for a vertical well placed between two intersecting sealing faults

where L is the distance between the well and the intersection point of the two sealing faults. Figure 2.10 show well testing diagnostic plot for a vertical well in a reservoir with intersecting sealing fault, displaying FRF.

2.6.5 Radial Composite Reservoir, Radial Flow in Outer Zone

Well test interpretation for radial composite reservoirs is quite difficult; particularly if a sealing fault is nearby (Fig. 2.11). Therefore, it is essential to understand the recorded dynamic pressure response affected by the nearby fault for more accurate well test interpretation. Such accurate analysis will help the reservoir engineers to make a more valuable field development plan. In case there are two regions of radial composite reservoir, it means that the rock properties of the inner reservoir region are not the same as the rock properties in the outer reservoir region. Typically, the inner reservoir region is defined as the formation damages region around the wellbore, and the outer reservoir region represents the undamaged reservoir region far from the wellbore (Fig. 2.12).

Figure 2.13 show the diagnostic plot for a radial composite reservoir, in a vertical well. The plot displays radial flow in the outer zone (RF2).

In the outer zone of an infinite circular composite reservoir, the pressure response during radial flow testing may be expressed as:

$$p_i - p_{wf} = \frac{162.6\, q\, B\mu}{Mkh} \left[\log\!\left(\frac{Mkt}{\emptyset \mu C_t r_w^2} \right) - 3.23 + 0.869\, (M_s + S_{RC}) \right] \quad (2.21)$$

where M is the mobility ratio of the outer region to that of the inner region, and k is the permeability of the inner zone, and the mobility ration can be expressed as:

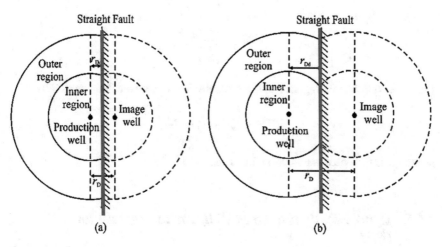

Fig. 2.11   Graphs of two different examples of a composite reservoir with a straight fault

Fig. 2.12   Show a
two-region radial composite
reservoir

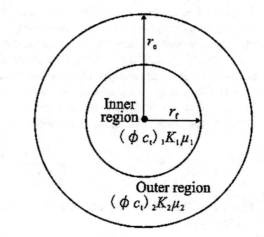

Fig. 2.13   Diagnostic plot in
a radial composite reservoir,
showing radial flow in the
outer zone (RF2)

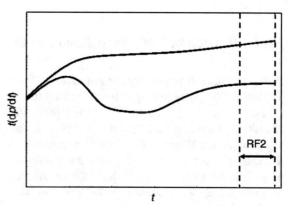

$$M = \frac{\left(\frac{K}{\mu}\right)_2}{\left(\frac{K}{\mu}\right)_1} \tag{2.22}$$

where the skin factor, S_{RC}, due to the pressure drop across the inner region is:

$$S_{RC} = (M - 1) \ln \frac{r_{e1}}{r_w} \tag{2.23}$$

where r_{e1} is the outer radius of the inner region.

2.6.6   Hydraulically Fractured Well, Pseudo-radial Flow (PRF)

Hydraulic fractured reservoirs increase production flow rates of oil and gas wells by making a conductive connection between the reservoir and the wellbore (Fig. 2.14). Normally, the well productivity of a fractured reservoir is depending on many parameters such as fracture dimensions, fracture conductivity, reservoir drainage area, reservoir conductivity, and type of formation damage generated through the process.

Figure 2.15 displays a diagnostic plot for a hydraulically fractured well displaying pseudo-radial flow (PRF). The recorded pressure response through the pseudo-radial flow with the overall skin factor is written as:

$$S = -\ln\left(\frac{1}{2 + \frac{1}{C_r}} \frac{L_f}{r_w}\right) \tag{2.24}$$

where s_f is the fracture damage skin factor and C_r is the dimensionless fracture conductivity.

2.6.7   Horizontal Well, Early Radial Flow

An early radial flow regime appears around every fracture after the linear flow regime. The radial flow phase mostly relies on fracture spacing and length. Also, during this type of flow regime, fractures still act independently (Fig. 2.16). Wellbore storage effects may hide data for ERF, but when present, they may be evaluated on a semi-log plot. In the absence of wellbore storage effects, the early radial flow might theoretically start at time zero. When the transient touches a vertical boundary or when flow occurs from beyond the wellbore, the end period of the early radial flow may happen. Typically, the smaller of these two values represents the completion of the flow time.

Fig. 2.14 Shows a hydraulic fracture reservoir with different flow regimes for a vertical well

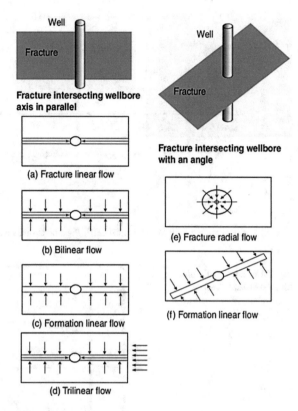

Fig. 2.15 Diagnostic plot for a hydraulically fractured well displaying PRF

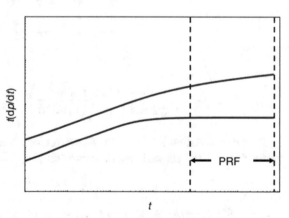

The pressure response equation for ERF to a horizontal well (see Fig. 2.17) can be written as:

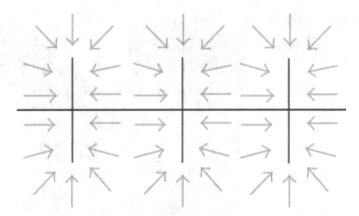

Fig. 2.16 Show early redial flow regime (ERF) for a horizontal well

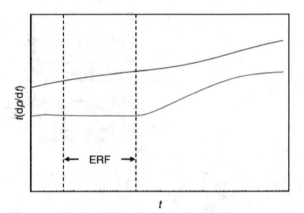

Fig. 2.17 Diagnostic plot
for a horizontal well
displaying ERF

$$p_i - p_{wf} = \frac{162.6\,q\,B\mu}{\sqrt{k_y k_z}\,L_h} \left[\log\left(\frac{\sqrt{k_y k_z}\,t}{\emptyset \mu C_t r_w^2} \right) - 3.23 + 0.869\,S \right] \tag{2.25}$$

By using equation Eq. 2.11 for IARF, replace h with wellbore length L_h, and permeability k by geometric mean permeability, $k_y\,k_z$.

2.6.8   Horizontal Well, Early Hemi-Radial Flow (EHRF)

Early hemi-radial flow occurs when the wellbore is very closer to a single boundary than the other boundaries, (Fig. 2.18). The early hemi-radial flow can appear exactly after the early radial flow regime. Eventually, the vicinity area influenced by the production will incorporate the whole thickness of the reservoir.

Fig. 2.18 Show a horizontal well displaying early hemi-radial flow

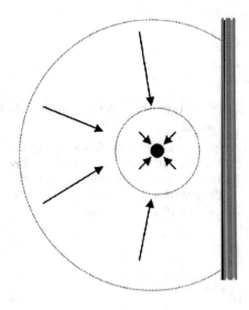

Figure 2.19 shows a diagnostic plot for early hemi-radial flow in a horizontal well. The pressure response equation for hemi-radial flow regime in a horizontal well exhibiting can be written as:

$$p_i - p_{wf} = \frac{325.3\,qB\mu}{\sqrt{k_y k_z}L_h}\left[\log\left(\frac{\sqrt{k_y k_z}t}{\emptyset\mu C_t r_w^2}\right) - 3.23\right] + \frac{141.2\,qB\mu}{\sqrt{k_y k_z}L_h}(S + S_{HRF})$$

$$(2.26)$$

where S_{HRF} is a negative geometric skin factor expressed as:

Fig. 2.19 Show diagnostic plot for early hemi-radial flow in a horizontal well

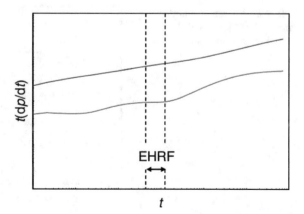

$$S_{HRF} = -\ln \frac{2d_z}{r_w} \tag{2.27}$$

2.6.9 Horizontal Well, Late Pseudo-radial Flow (LPRF)

A type of radial flow that takes place afterwards is known as pseudo-radial flow. The pseudo-radial flow will persist during the prolonged test for fractured wells if the fracture half-length is not particularly long, as demonstrated in Figs. 2.20 and 2.21.

The governing equation for the pressure response for late-pseudo-radial flow in a horizontal well can be written as:

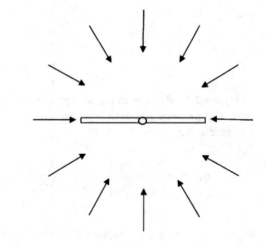

Fig. 2.20 Show Late pseudo-radial flow starts once flow enters the wellbore from the outside ends of the well

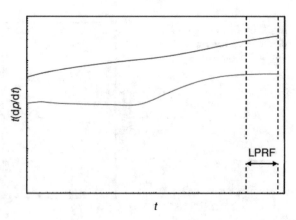

Fig. 2.21 Show a diagnostic plot for late-pseudo-radial flow in a horizontal well

$$p_i - p_{wf} = \frac{162.6\,q\,B\mu}{\sqrt{k_x k_h}\,h}\left[\log\left(\frac{\sqrt{k_x k_y}\,t}{\emptyset\mu C_t r_w^2}\right) - 3.23 + 0.869\,S_t\right] \qquad (2.28)$$

where S_t is the overall skin factor.

2.7 Linear Flow

In a linear flow regime, the fluid flows linearly from the reservoir to the fractures and every single fracture acts independently (Fig. 2.22). Linear flow during fracture controls early time data. Linear flow is also appearing in channel reservoirs and horizontal wells. If permeability is provided, the collected data from the linear flow time could be applied to calculate channel width or fracture half-length. If the productive well length open to flow is known for horizontal wells, it is possible to determine the permeability perpendicular to the well.

2.7.1 Linear Flow Caused by Hydraulically Created Fracture

Usually, when a well is hydraulically fractured, the bottom hole causes a single vertical fracture that intersects the well. The reservoir thickness is nearly equal to the height of the fracture, which has a half-length of xf. A fracture with "infinite conductivity" has a permeability that is often quite high. Figure 2.23 illustrates the linear flow.

Figure 2.24 displays linear flow on the diagnostic plot when a derivative follows a half-slope line, or a line that travels up two log cycles in the vertical direction for every one log cycle in the horizontal direction. The change in pressure might also follow a half-slope line or not. If the fracture is undamaged, a half-slope line will be followed by the pressure change in a hydraulically fractured well. The pressure

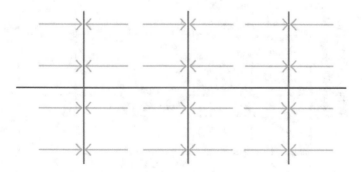

Fig. 2.22 Show the liner flow regime for a multi-staged fractured horizontal well

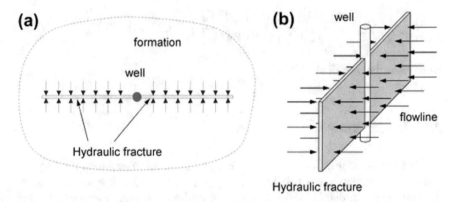

(a)

formation

well

Hydraulic fracture

(b)

well

flowline

Hydraulic fracture

Fig. 2.23 Linear flow near a fractured well. **a** Plane view **b** Longitudinal

changes in a channel formation, a horizontal well, or a hydraulically fractured well with damage, will move toward the half-slope line from above. The logarithmic derivative plot shows the period of linear flow has a slope of one-half, as exhibited in Figs. 2.25, 2.26, 2.27, 2.28 and 2.29.

The governing equation of the pressure response during linear flow is a linear function of the square root of time, which is expressed as:

$$\Delta p = m_L \sqrt{t} + b_L \qquad (2.29)$$

On a plot of pressure versus the square root of time, the pressure response data for linear flow is represented by a straight line with a slope of m_L and an intercept of b_L. The slope m_L and cross-sectional area exposed to the flow are inversely related to the flow pathway's permeability. If the linear flow is the first to arise and there are no flow constraints, the curve will have a slope of one-half, as shown in Fig. 2.27.

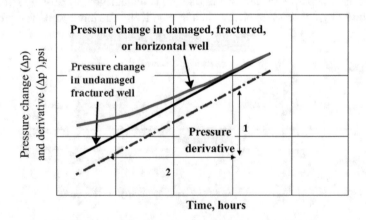

Fig. 2.24 Linear flow derivative follows a half-slope line on a diagnostic plot

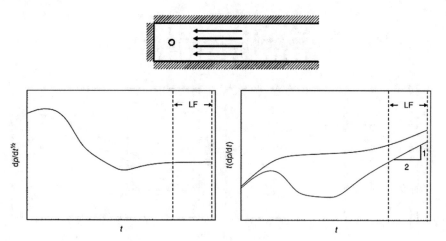

Fig. 2.25 Show linear flow in one direction for a well located in a channel with one closed end

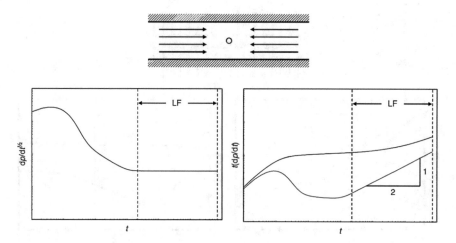

Fig. 2.26 Show the linear flow in two directions for a well located in a channel

The curve will be twice as high as the derivative curve. If there is a constraint to flow, such as in a choked fracture, or if other flow types occur before the linear flow regime, as in the case of a well placed in a channel, the curve will virtually resemble a straight line with a slope of one-half, as seen in Figs. 2.25, 2.26 and 2.29.

Typically, the linear flow diagnostic plot is a log–log plot of the linear flow derivative, $dp/dt^{1/2}$ versus time. The derivative of linear flow can be determined from the logarithmic derivative (Stewart 2011, p. 389):

$$\frac{dp}{dt^{1/2}} = \frac{2}{t^{\frac{1}{2}}}\left(t\frac{dp}{dt}\right) \tag{2.30}$$

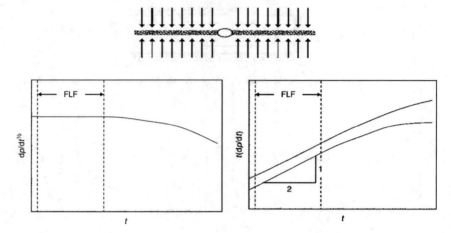

Fig. 2.27 Show the linear flow for a well with high-conductivity vertical hydraulics

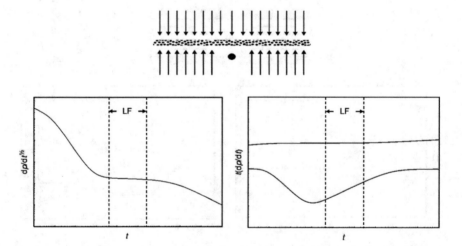

Fig. 2.28 Show the linear flow for a well near a single high conductivity fault

If the cross-section area of the flow route is identified, the slope m_L can be applied to determine the permeability in the flow direction. The slope m_L can be determined from the straight-line section of a plot of pressure versus $t^{1/2}$. The following equation can be used to estimate the slope m_L using the field data derivative of the logarithmic diagnostic plot:

$$m_L = \frac{2}{t_L^{\frac{1}{2}}}\left(t\frac{dp}{dt}\right)_L \qquad (2.31)$$

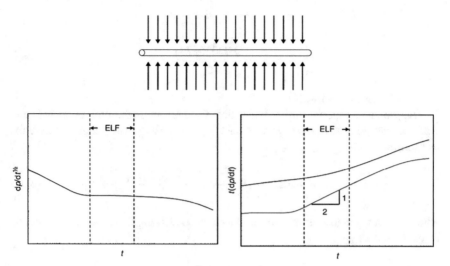

Fig. 2.29 Show linear flow for a Horizontal well

where $(t\, dp/dt)_L$ and t_L are the logarithmic derivative and time coordinates, respectively, for any point on the half-slope line related to linear flow. Instead, the slope, m_L, can be read from the horizontal section of the derivative on the linear flow diagnostic plot.

Several reservoir models may include linear flow. Linear flow will be discussed in the following sections for a vertical well positioned in a channel with one or both ends open, a hydraulically fractured well, and a horizontal well displaying early linear flow.

2.7.2   Channel Reservoir, Both Ends Open

Figure 2.26 shows the well that is placed in the channel reservoir with both open ends. A well located in an infinite channel with linear flow in both directions is shown in the plot illustrating pressure and pressure derivative response, along with the diagnostic chart for linear flow.

The governing equation for dimensionless pressure response for linear flow along both open ends channel can be written as (Ehlig-Economides and Economides 1985):

$$p_D = 2\sqrt{\pi t_{wD}} \tag{2.32}$$

where t_{wD} and p_D are identified as:

$$p_D = \frac{kh}{141.2\,qB\mu}\left(p_i - p_{wf}\right) \tag{2.33}$$

And

$$t_{wD} = \frac{0.0002637 \, kt}{\emptyset \mu C_t W^2} \tag{2.34}$$

where w is the channel width.

The pressure response during linear flow for a channel with both open ends and with incorporating skin factor and geometric effects can be written in oilfield units as follows:

$$p_i - p_{wf} = 8.128 \frac{qB}{hw} \sqrt{\frac{\mu t}{k \emptyset C_t}} + \frac{141.2 \, q B \mu}{kh} (S + S_c) + \tag{2.35}$$

where S_c is a geometric skin factor produced by converging flow. The skin factor can be expressed as:

$$S_c = \ln \frac{w}{r_w} - \ln \sin \frac{\pi d}{w} - 1.838 \tag{2.36}$$

where d is the distance from the well location to one edge of the channel.

If the reservoir permeability is identified, the width of the channel can be determined from the slope m_L using the following equation:

$$w = 8.128 \frac{qB}{h|m_L|} \sqrt{\frac{\mu}{k \emptyset C_t}} \tag{2.37}$$

2.7.3   Channel Reservoir, One End Open

Figure 2.25 illustrates the well located in the channel reservoir closed at one side. The plot displaying pressure and pressure derivative response, together with the linear flow diagnostic graph, for a well placed in semi-infinite channel with the linear flow in one direction along the channel.

The governing equation for pressure response for the linear flow along one side open ended channel can be written as:

$$\Delta p = 16.256 \frac{qB}{hw} \sqrt{\frac{\mu t}{k \emptyset C_t}} \tag{2.38}$$

If the reservoir permeability is identified, the channel width could be determined using the data in a linear flow period along one flow direction in a channel. The width of the channel can be determined using the following equation:

$$w = 16.256 \frac{qB}{h|m_L|} \sqrt{\frac{\mu}{k\emptyset C_t}} \tag{2.39}$$

2.7.4 High-Conductivity Hydraulic Fracture, Formation Linear Flow

The well with a vertical hydraulic fracture that has a high conductivity is shown in Fig. 2.27, along with the linear flow of the reservoir. Plot illustrating reservoir linear flow for a well located in a high-conductivity vertical fracture, together with pressure and pressure derivative response and a diagnostic graph for linear flow.

According to Cinco-Ley and Samaniego (1981), the following formula may be used to represent the pressure response for a well with a high-conductivity vertical fracture and a reservoir with a linear flow regime:

$$p_i - p_{wf} = 4.064 \frac{qB}{hL_f} \sqrt{\frac{\mu t}{k\emptyset C_t}} + \frac{141.2\, qB\mu}{kh} \left(S_f + S_{fcf}\right) \tag{2.40}$$

where S_f denoted the fracture damage skin factor, S_{fcf} denoted the apparent skin produced by the finite fracture conductivity, and L_f denoted the fracture half-length, (Camacho et al. 1987). The apparent skin can be determined using the following equation:

$$S_{fcf} = \frac{1}{3C_r} \tag{2.41}$$

where C_r is the dimensionless fracture conductivity, which can be expressed as:

$$C_r = \frac{w_f k_f}{\pi k L_f} \tag{2.42}$$

If the reservoir permeability is identified, the fracture half-length, L_f, can be determined using the following equation:

$$L_f = 4.064 \frac{qB}{hm_L} \sqrt{\frac{\mu t}{k\emptyset C_t}} \tag{2.43}$$

2.7.5   Well Near a High-Conductivity Fault

Figure 2.28 shows reservoir linear flow in a well located close to the center of a single high-conductivity fault (Cinco-Ley et al. 1976). The plot shows the diagnostic graph for linear flow together with the pressure and pressure derivative response. This flow model is equivalent to a well with a high-conductivity vertical fracture model with choked fracture skin S_f.

The governing equation for pressure response for a well with a high-conductivity vertical fracture can be written as:

$$p_i - p_{wf} = 4.064 \frac{qB}{hL_f} \sqrt{\frac{\mu t}{k \varnothing C_t}} + \frac{141.2\,qB\mu}{kh} S_f \tag{2.44}$$

In case the well lies on the perpendicular bisector of the fault, the S_f can be determined using the following equation:

$$S_f = \ln \frac{2L}{r_w} \tag{2.45}$$

where L is the distance from the location of the well to the high-conductivity fault.

2.7.6   Horizontal Well, Early Linear Flow

Figure 2.29 illustrates a horizontal well displaying early linear flow. The plot displays pressure and pressure derivative response, together with the linear flow diagnostic graph.

The governing equation for pressure response for a horizontal well showing early linear flow can be written as:

$$p_i - p_{wf} = \frac{8.128\,qB}{L_h h} \sqrt{\frac{\mu t}{k_y \varnothing C_t}} + \frac{141.2\,qB\mu}{\sqrt{k_y k_z} L_h} (S + S_A + S_C) \tag{2.46}$$

where S_C is a positive geometric skin factor produced by converging flow, which may express as:

$$S_c = \ln \frac{h}{r_w} + 0.25 \ln \frac{k_y}{k_z} - \ln \sin \frac{\pi d_z}{h} - 1.838 \tag{2.47}$$

and S_A is a negative geometric skin factor produced by permeability anisotropy, which may express as:

$$S_A = \ln \left[\frac{1}{2} \left(\left(\frac{k_y}{k_z} \right)^{1/4} \right) + \left(\frac{k_z}{k_y} \right)^{1/4} \right] \tag{2.48}$$

2.7.7 Example 2.1: (Combined Analyzing of Radial and Linear Flow)

Well -Y is considered to be at the centre of the fluvial channel. The well was tested to a 72-h drawdown test. Tables 2.3 and 2.4 contain the test results. Identify and assess data demonstrating radial and linear flow.

Solution

1. Radial and linear flow regimes have previously been distinguished from the log–log diagnostic plot shown in Fig. 2.30. The horizontal part of the derivative shows that the radial flow phase lasts for approximately 7 h and begins around an hour later. The linear flow phase begins around 14 h into the test and lasts until the end.
2. Draw a horizontal line over the data in the radial flow regime and interpret the value of the derivative during IARF, $(t \, \Delta p')$, as 6.9 psi.
3. Determine the permeability from the data in IARF using the below equation:

$$k = \frac{70.6 \; q B \mu}{h (t \Delta p')_r}$$

$$k = \frac{70.6 \times 125 \times 1.18 \times 1.06}{23 \times 6.9} = 70 \, \text{mD}$$

4. Find the starting and end of the linear flow phase on the square root of the time plot using the following equation:

$$t_{bLF}^{1/2} = \sqrt{4}$$

Table 2.3 Shows rock and fluid property data for radial/linear flow

q	125 STB/D
h	23 ft
ϕ	18%
p_i	1930 psi
r_w	0.25 ft
B_o	1.18 bbl/STB
μ_o	1.06 cp
C_t	$17.7 \times 10^{-6} \, \text{psi}^{-1}$

Table 2.4 Show test data for radial/linear flow analysis

t(h)	P_{wf} (psia)	t(h)	P_{wf} (psia)	t(h)	P_{wf} (psia)	t(h)	P_{wf} (psia)
0.0050	1915.78	0.256	1799.43	1.910	1783.64	12.80	1770.46
0.0106	1902.13	0.293	1797.77	2.154	1782.78	14.40	1769.40
0.0170	1888.64	0.335	1796.77	2.428	1782.06	16.21	1768.32
0.0241	1875.96	0.382	1795.37	2.737	1781.18	18.24	1767.14
0.0321	1864.29	0.435	1794.52	3.084	1780.65	20.53	1765.98
0.0411	1853.57	0.494	1793.3	3.474	1779.54	23.10	1764.75
0.0512	1843.77	0.561	1792.26	3.914	1778.93	25.99	1763.21
0.0626	1835.07	0.636	1791.22	4.408	1778.21	29.24	1761.86
0.0755	1827.67	0.720	1790.33	4.964	1777.36	32.90	1760.29
0.0899	1821.33	0.815	1789.49	5.589	1776.6	37.02	1758.76
0.1061	1815.91	0.922	1788.81	6.293	1775.71	41.65	1757.20
0.1244	1811.71	1.042	1787.79	7.085	1774.85	46.86	1755.29
0.1449	1808.05	1.178	1786.91	7.975	1774.08	52.73	1753.34
0.1681	1805.05	1.330	1786.03	8.977	1773.25	59.32	1751.19
0.1941	1802.93	1.501	1785.14	10.10	1772.31	66.74	1749.15
0.223	1800.97	1.694	1784.61	11.37	1771.28	72.00	1747.51

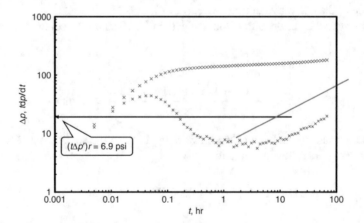

Fig. 2.30 Log–log diagnostic plot, radial and linear flow

$$t_{bLF}^{1/2} = 3.7 \ h^{1/2}$$

and

$$t_{eLF}^{1/2} = \sqrt{72}$$
$$t_{eLF}^{1/2} = 8.5 \ h^{1/2}$$

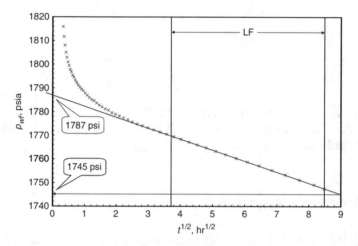

Fig. 2.31 Square-root of time plot, radial and linear flow

5. Plot a straight line over the test data between 3.7 $h^{1/2}$ and 8.5 $h^{1/2}$ on the square root of the time plot, Fig. 2.31.

6. Determine the linear flow slope m_L from two test points on the straight line using the following slope equation:

$$|m_L| = \left| \frac{1787 - 1745}{9 - 0} \right| = 4.67 \ h^{1/2}$$

7. Determine the channel width from the following equation:

$$w = 8.128 \ \frac{qB}{h|m_L|} \sqrt{\frac{\mu}{k\emptyset C_t}}$$

$$w = 8.128 \times \frac{125 \times 1.18}{23 \times 4.67} \sqrt{\frac{1.06}{70 \times 0.18 \times 17.7 \times 10^{-6}}} = 770 \ ft$$

8. Determine the radius of investigation at the start and end of the radial flow phase, using the following equation:

$$r_{ibrf} = \sqrt{\frac{kt_{bf}}{948\emptyset\mu C_t}}$$

$$r_{ibrf} = \sqrt{\frac{70 \times 1}{948 \times 0.18 \times 1.06 \times 17.7 \times 10^{-6}}} = 147 \ ft$$

Similarly,

$$r_{ibrf} = \sqrt{\frac{kt_{bf}}{948\emptyset\mu C_t}}$$

$$r_{ierf} = \sqrt{\frac{70 \times 7}{948 \times 0.18 \times 1.06 \times 17.7 \times 10^{-6}}} = 390\,\text{ft}$$

As the well is located in the channel, the distance between the well position and either side of the channel is 385 feet, or half the width, which is nearly identical to the predicted investigational radius.

2.8   Volumetric Behaviour

Through volumetric behaviour, the pressure response is a linear function of time. Therefore, it is defined as that pressure response time dominated by the wellbore or reservoir, acting like equal pressure "as a tank" with fluid entering or exiting the tank. The wellbore storage (WBS) is a good example of volumetric behaviour, which dominates for the duration of the early-time region. The wellbore behaves like a tank with constant pressure. Fluid either exists WBS (initial times in a flow test before the reservoir starts to respond) or enters the WBS (earliest times in a BU test).

This pressure has the following linear equation form:

$$\Delta p = m_v t + b_v \tag{2.49}$$

The plotted pressure data shows the volumetric behaviour as a straight line with slope m_V and intercept b_V on a plot of pressure versus time.

The volumetric behaviour has a unit-slop line when plotting the logarithmic derivative, as shown in Fig. 2.32. As volumetric behaviour is the first flow regime to occur in WBS, the pressure change curve will also show a unit-slope line, as exhibited in Fig. 2.33. Throughout other flow periods showing volumetric behaviour, the pressure change curve will approximately move toward a unit-slope line from above, as displayed in Figs. 2.32, 2.34 and 2.35.

The slope m_V could be utilized to determine the volume creating the volumetric behaviour. The m_V can be obtained from the straight-line section of a plot of pressure versus time. Also, it can be determined from the field data derivative of the log–log diagnostic plot using the following equation:

$$m_v = \frac{1}{t_v}\left(t\frac{dp}{dt}\right)_v \tag{2.50}$$

where $(tdp/dt)_V$ is the logarithmic derivative and t_V is the time, for any data points that fall on the unit-slope line related to volumetric behaviour. Instead, the slope can be read from the horizontal line of the derivative on the volumetric diagnostic plot. Normally, the volumetric behaviour involves WBS and PSSF, along with other flow

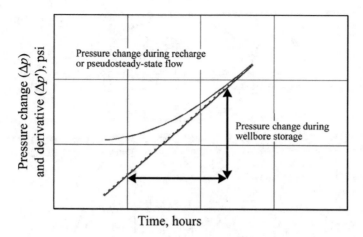

Fig. 2.32 Volumetric flow creates derivative with unit-slope line

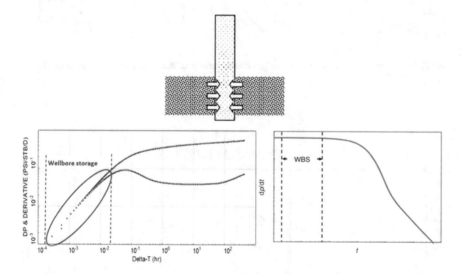

Fig. 2.33 Vertical well test showing constant wellbore storage

regimes. In the next sections, volumetric behaviour during WBS, PSSF in a closed reservoir, and in a radial composite reservoir when the mobility of the inner region is much higher than that of the outer region will be discussed in more detail.

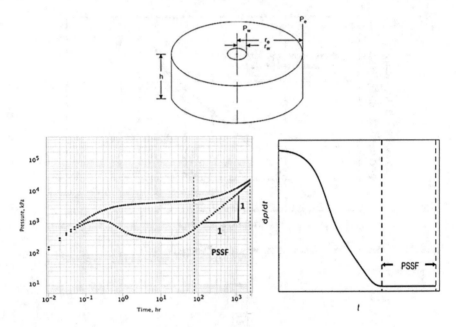

Fig. 2.34 Vertical well located in a closed circular reservoir, showing PSSF through a drawdown test

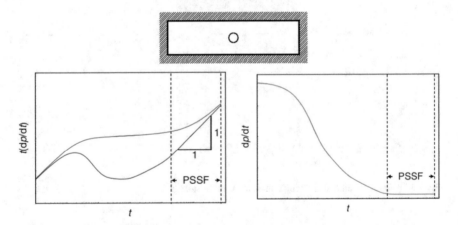

Fig. 2.35 Well located in closed rectangular reservoir, showing PSSF through a drawdown test

2.9   Phenomenon of Wellbore Storage

WBS is a phenomenon of phase redistribution that happens in a well that is shut in at the surface choke valve or at the bottom hole valve with gas and liquid flowing at the same time into the wellbore from the reservoir. WBS effect is identified directly after a production start-up or shut-in when the pressure behaviour at early times is controlled by the compressibility and volume of the wellbore fluid. WBS is a function of the wellbore fluid and the complete volume. Wellbore Storage is already discussed in Chap. 1 and Sect. 2.10 in more detail.

2.10 Closed Reservoir, Pseudo-steady State Flow (PSSF)

Pseudo-steady state flow during a constant rate drawdown is a particular case of volumetric behaviour, Figs. 2.32 and 2.33.

Figure 2.32, which depicts the pressure behaviour during PSSF from a closed circular reservoir, can be expressed as follows:

$$p_{wf}(t) = p_i - \frac{0.0744\,qBt}{r_e^2 h \emptyset C_t} - \frac{141.2\,qB\mu}{kh}\left[\ln\left(\frac{r_e}{r_w}\right) - \frac{3}{4} + S\right] \tag{2.51}$$

where r_w is the wellbore radius and r_e is the drainage radius of the reservoir. The following equation is used for pseudo-steady state flow for a closed reservoir:

$$p_{wf}(t) = p_i - \frac{0.234\,qBt}{Ah\emptyset c_t} - \frac{141.2\,qB\mu}{kh}\left[\frac{1}{2}\ln\left(\frac{10.06\,A}{C_A r_w^2}\right) - \frac{3}{4} + S\right] \tag{2.52}$$

where A is the drainage area of the reservoir, and C_A is a drainage area shape factor which depends on the reservoir shape and the well location in the reservoir.

The slope, m_V, for the period of pseudo-steady state flow is inversely proportional to the volume drained and is independent of the reservoir form. The reservoir pore volume V_p can be determined from the slope throughout a reservoir limits test using the following equation:

$$V_p = Ah\emptyset = \frac{0.234\,qB}{|m_v|C_t} \tag{2.53}$$

Initial oil in place can be estimated using the following equation:

$$N = \frac{q(1 - S_w)}{24|m_{pss}|C_t} \tag{2.54}$$

2.11 Radial Composite Reservoir With Low Mobility Outer Zone

A well situated in a radial composite reservoir, where the outer part has significantly lower mobility than the inner portion (Fig. 2.36), may exhibit volumetric
behaviour (VB) for both buildup and drawdown tests, in comparison to the buildup
test behaviour for a well positioned in a closed reservoir. Because fluid entry from
the outer layer with reduced mobility recharges the inner layer with greater mobility
during the BU test, this behaviour is known as recharge.

2.11.1 Example 2.2

Tables 2.5 and 2.6 are the obtained from the reservoir limits test data. Analyse the
test data.

Solution

Fig. 2.36 Vertical well
located in an infinite radial
composite reservoir, showing
volumetric behaviour

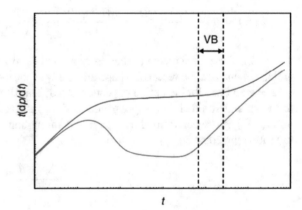

Table 2.5 Show rock and
fluid property data

q	250 STB/D
h	15 ft
ϕ	21%
S_w	25%
r_w	0.32 ft
B_o	1.328 bbl/STB
μ_o	0.61 cp
Ct	16.1×10^{-6} psi^{-1}

Table 2.6 Show drawdown test data for reservoir limits

t (h)	P_{ws} (psia)	t (h)	P_{ws} (psia)	t (h)	P_{ws} (psia)	t (h)	P_{ws} (psia)
0	4419.0	5	4209.6	18	4203.4	45	4193.4
0.5	4240.8	6	4208.3	20	4202.5	48	4192.5
0.75	4225.4	7	4207.4	22	4202.1	51	4190.8
1	4219.5	8	4207.3	24	4201.2	54	4189.9
1.25	4217.0	9	4207.0	27	4199.7	57	4188.5
1.5	4215.4	10	4206.3	30	4198.4	60	4187.8
2	4213.3	11	4206.2	33	4197.7	63	4186.9
2.5	4212.2	12	4205. 7	36	4196.8	66	4185.2
3	4211.4	14	4204.9	39	4195.4	69	4184.1
4	4210.2	16	4203.7	42	4194.5	72	4183.8

1. Plot bottomhole pressure, p_{wf}, versus time, t, on a Cartesian scale, as exhibited in Fig. 2.37
2. Plot a straight line through the test data in PSSF period.
3. Determine the slope m_{pss} and the intercept b_{pss} of the straight line. The intercept b_{ps} from the plot is 4210.0 psi, and the slope is given by:

$$m_{pss} = \frac{4210.0 - 4180.3}{0 - 80} = 0.37 \text{ psi/h}$$

4. Determine the reservoir pore volume, V_p, and the oil in place, N, from the slope m_{pss} as:

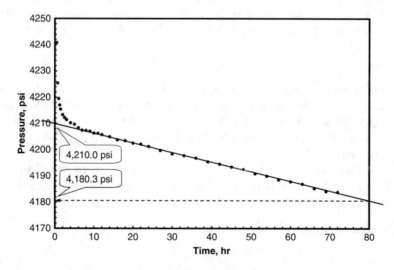

Fig. 2.37 Reservoir limits DD test analysis

$$V_p = \frac{0.234 \times 250 \times 1.328}{0.371 \times 16.1 \times 10^{-6}} = 13 \times 10^6 \text{ ft}^3$$

Oil initial in place can be estimated using the following equation:

$$N = \frac{q \ (1 - S_w)}{24 |m_{pss}| C_t}$$

$$N = \frac{250 \times \ (1 - 0.25)}{24 \ \times \ 0.371 \times 16.1 \times 10^{-6}} = 1.31 \times 10^6 \text{ STB}$$

5. Calculate the drainage area by known thickness net pay and porosity:

$$A = \frac{V_p}{\emptyset h} = \frac{13 \times 10^6}{0.21 \times 15} = 4.13 \times 10^6 \text{ ft}^2 \approx 94.7 \text{ acre.}$$

6. Calculate the productivity index, J, by known reservoir pressure and intercept b_{pss}:

$$J = \frac{q}{p_i - b_{pss}} = \frac{250}{4419 - 4210} = 1.20 \text{ STB/D/Psi}$$

2.11.2   Example 2.3

Use the data in example 2.2 (Tables 2.5 and 2.6) obtained from the drawdown test of well-K102. Identify and analyze data in pseudo-steady state flow using the log–log and primary pressure derivative diagnostic plots. Confirm that the correct part of the data was analyzed as pseudo-steady state flow in Fig. 2.38.

Solution

1. From the log–log diagnostic plot exhibited in Fig. 2.39, the volumetric behaviour (PSSF) from the final unit-slope line has been identified. The PSSF phase starts around 12 h and continues through the end of the test.
2. Select a point on the pseudo-steady-state unit-slope line and read $(t \Delta p')_V = 37$ psi at time $t_V = 100$ h.
   Another option, the slope can be read from Fig. 2.40 from the horizontal part of the primary pressure derivative. The plot shows that PSSF starts around 12 h and continues through the end of the test at 72 h.

3. Estimate the volumetric slope as follows:

$$m_v = \frac{1}{t_v} \left(t \frac{dp}{dt} \right)_v$$

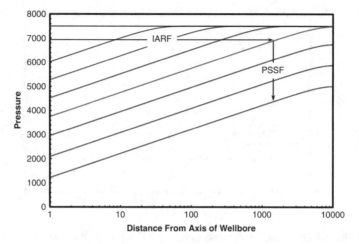

Fig. 2.38 Reservoir pressure profiles for IARF and PSSF

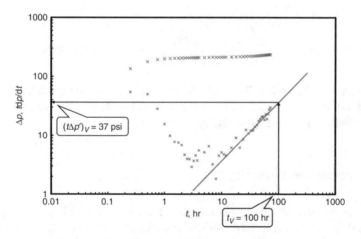

Fig. 2.39 Estimating slope from the log–log diagnostic plot, PSSF phase

$$m_v = \frac{37}{100} = 0.37 \ \text{psi/h}$$

4. Estimate the reservoir pore volume, Vp, and the oil in place, N:

$$V_p = \frac{0.234 \times 250 \times 1.328}{0.371 \times 16.1 \times 10^{-6}} = 13 \times 10^6 \ \text{ft}^3$$

Oil initial in place can be estimated using the following equation:

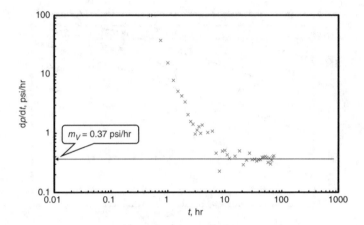

Fig. 2.40 Estimating the slope from primary pressure derivative plot, PSSF phase

$$N = \frac{q\,(1 - S_w)}{24\,|m_{pss}|C_t}$$

$$N = \frac{250 \times\ (1 - 0.25)}{24 \times\ 0.371 \times\ 16.1 \times 10^{-6}} = 1.31 \times 10^6 \ \text{STB}$$

5. In Fig. 2.38 the straight line for PSSF analysis goes through the data recognized as PSSF in Fig. 2.39, from 12 h through the end of the test. So, the accurate data was utilized for the analysis in Fig. 2.38.

2.12 Spherical Flow Regime

When the dominating flow pattern in the reservoir is directed toward a point, a spherical flow regime is formed. For well completions with limited entry and partial penetration, this flow occurs (Chatas 1966). The pressure derivative on the log–log diagnostic plot identifies this flow regime as a half-slope line. Its occurrence makes it possible to calculate the spherical permeability. Calculations of both vertical and horizontal permeabilities are possible after the spherical flow regime and before the radial flow domain.

$$\Delta p = m_s t^{-1/2} + b_s \tag{2.55}$$

On a plot of pressure versus $t^{1/2}$, data demonstrating spherical flow will follow a line with a slope of m_S and an intercept of b_S. As seen in Fig. 2.41, during spherical flow, the pressure change curve approaches a constant value while the log–log derivative plot has a slope of minus one-half.

The spherical flow derivative can be determined from the logarithmic derivative using the following equation:

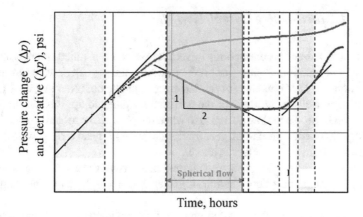

Fig. 2.41 The spherical flow derivative has the slope of $-1/2$

$$\frac{dp}{dt^{-1/2}} = 2t^{1/2}\left(t\frac{dp}{dt}\right) \tag{2.56}$$

The slope m_S can be calculated from the slope of a plot of p versus $1/t^{1/2}$. The spherical flow slope can also be calculated from the log–log diagnostic plot using the following equation:

$$m_s = 2t_s^{1/2}\left(t\frac{dp}{dt}\right)_s \tag{2.57}$$

where $(tdp/dt)_S$ is the pressure derivative and t_s is the time of any data point on the negative half-slope line related to the spherical flow period. Also, m_S can be read from the horizontal section of the test data on the spherical flow diagnostic graph. Both Proett and Chin (1998) identify the spherical derivative as:

$$t^{3/2}\frac{dp}{dt} = \frac{1}{2}\frac{dp}{dt^{-1/2}} = \frac{1}{2}m_s \tag{2.58}$$

The spherical flow slope m_S can be applied to determine the spherical permeability, k_s, using the following equation:

$$k = \sqrt[3]{k_x k_y k_z} \tag{2.59}$$

In the next sections, pressure responses a well with a limited-entry completion and for a well with partial penetration will be discussed.

2.12.1   Limited-Entry Completion

Normally, limited-entry completions in vertical wells are planned to prevent unde-
sirable fluid production, such as gas production from the gas cap or water production
from the aquifer. The influences of limited entry could be observed in gravel-packed
and perforated wells when some of the perforations plugged up. It is also known as
partial completion. Such flow occurs also when the productive reservoir zone is only
partially penetrated. The partial penetration describes a near-well flow restriction
that causes a positive skin effect in a well.

Figure 2.42 exhibits the pressure and pressure derivative response for a vertical
well with a limited-entry completion, together with the spherical flow diagnostic
plot.

The pressure response for a limited-entry completion is expressed by the following
equation:

$$\Delta p = \frac{70.6 B \mu}{k_s r_s} \left(1 + \frac{2 r_s}{h_p}\right) - 2453 \frac{q B \mu}{k_s} \sqrt{\frac{\emptyset \mu C_t}{k_s t}} \qquad (2.60)$$

where h_p is the height of the perforated interval, k_s is the spherical permeability
which can be calculated using Eq. 2.59. Alternatively, the spherical permeability
could be determined from the slope m_S of test data showing the spherical flow period
and apply the following equation:

$$k_s = \left(\frac{2453 \, q B \mu \sqrt{\emptyset \mu c_t}}{m_s}\right)^{2/3} \qquad (2.61)$$

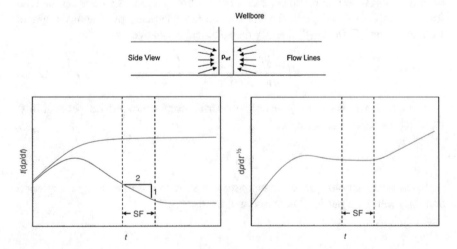

Fig. 2.42 Vertical well with limited-entry completion, exhibiting spherical flow

The horizontal permeability, k_h, can be calculated using the following equation:

$$k_h = \sqrt{k_x k_y} \qquad (2.62)$$

The equivalent spherical radius can be determined using the following equation:

$$r_s = \sqrt{r_w h_p} \qquad (2.63)$$

2.12.2 Partial Penetration

Typically, during partial penetration completion, there are three types of flow regime geometries are seen during the fluid flow in the reservoir: radial flow around the producing well section, spherical flow, and redial flow across the entire reservoir width. Partial penetration and end effects usually complicate interpretation.

Figure 2.43 illustrates the pressure and pressure derivative response, with the spherical flow diagnostic graph, for partial penetration completion well.

The governing pressure response equation for hemispherical flow to a partial-penetration completion can be written as follows:

$$\Delta p = \frac{141.2\,q B \mu}{k_s r_s}\left(1 + \frac{r_s}{h_p}S\right) - 4906\frac{q B \mu}{k_s}\sqrt{\frac{\emptyset \mu C_t}{k_s t}} \qquad (2.64)$$

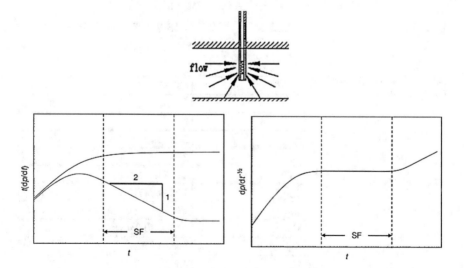

Fig. 2.43 Vertical well with partially penetrating completion, showing spherical flow regime

The spherical permeability, k_s, can be calculated from test data showing hemispherical flow from the following equation:

$$k_s = \left(\frac{4906 \, q \, B \mu \sqrt{\varnothing \mu c_t}}{m_s} \right)^{2/3} \tag{2.65}$$

And the equivalent spherical radius, r_s, for a partial penetration completion can be calculated using the following equation:

$$r_s = \sqrt{2 r_w h_p} \tag{2.66}$$

2.12.3 Example 2.4

Well-R22 is partially completed in the top 15 feet of a 100-foot reservoir thickness. A 12-h drawdown test was performed. Determine and assess the test data indicating spherical and radial flow based on the data in Tables 2.7 and 2.8.

Solution

1. Determine the total compressibility, c_t, using the following equation:

$$C_t = C_f + S_o C_o + S_g C_g + S_w C_w$$
$$C_t = 4 \times 10^{-6} + (1 - 0.22) \times \left(10.5 \times 10^{-6} \right)$$
$$+ 0 + 0.22 \times 4 \times 10^{-6} = 13 \times 10^{-6} \; \text{psi}^{-1}$$

Table 2.7 Show rock and fluid property data

q	625 STB/D
h	100 ft
ϕ	27%
p_i	2735 psi
S_w	22%
r_w	0.325 ft
B_o	1.21 bbl/STB
μ_o	1.06 cp
C_o	$10.5 \times 10^{-6} \; \text{psi}^{-1}$
C_w	$3 \times 10^{-6} \; \text{psi}^{-1}$
C_f	$4 \times 10^{-6} \; \text{psi}^{-1}$

Table 2.8 Show DD test data for reservoir limits

t (h)	P_{ws} (psia)	t (h)	P_{ws} (psia)	t (h)	P_{ws} (psia)	t (h)	P_{ws} (psia)
0.0014	2704.62	0.0625	2516.53	0.420	2490.34	2.514	2481.09
0.0030	2677.67	0.0718	2513.37	0.474	2489.44	2.829	2480.62
0.0047	2653.35	0.0821	2510.61	0.535	2488.68	3.184	2480.15
0.0067	2631.68	0.0938	2508.06	0.603	2487.79	3.584	2479.73
0.0090	2612.53	0.1069	2505.76	0.680	2487.14	4.033	2479.41
0.0115	2595.76	0.1217	2503.78	0.766	2486.38	4.539	2478.87
0.0143	2581.19	0.1383	2501.94	0.864	2485.83	5.107	2478.39
0.0175	2568.61	0.1570	2500.10	0.973	2485.22	5.747	2477.93
0.0211	2557.95	0.1780	2498.62	1.096	2484.7	6.467	2477.53
0.0252	2548.81	0.202	2497.15	1.234	2484.06	7.277	2477.04
0.0297	2541.18	0.228	2495.82	1.390	2483.47	8.188	2476.61
0.0348	2534.57	0.258	2494.59	1.565	2482.99	9.213	2476.14
0.0406	2529.05	0.292	2493.35	1.762	2482.44	10.37	2475.71
0.0471	2524.21	0.330	2492.37	1.984	2482.07	11.66	2475.29
0.0543	2520.18	0.372	2491.30	2.233	2481.54	12	2475.12

2. Plot test data on log–log diagnostic as exhibited in Fig. 2.44 and identify both spherical and radial flow periods. The derivative's negative half-slope line identifies the spherical flow phase, which begins at around 0.2 h and finishes at about one hour. The radial flow phase begins about two hours into the test period and lasts until the end.

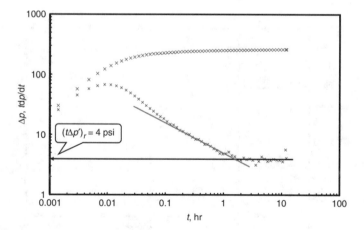

Fig. 2.44 Show Log–log diagnostic graph, showing spherical/radial flow regimes

3. Draw a horizontal line through the test data in the radial flow period and read the $(t\,\Delta p'')_r$ value during the IARF, as 4 psi.
4. Estimate the permeability from the test data in the IARF period using the following equation:

$$k = \frac{70.6\,q\,B\mu}{h(t\,\Delta p')_r} = \frac{70.6 \times 625 \times 1.21 \times 1.06}{100 \times 4} = 141\,\text{mD}.$$

5. Find the spherical flow period, starting and end, on the inverse square root of time plot:

$$t_{bSF}^{-1/2} = \frac{1}{\sqrt{0.2}} = 2.24\,\text{h}^{-1/2}$$

and

$$t_{eSF}^{-1/2} = \frac{1}{\sqrt{1}} = 1\,\text{h}^{-1/2}$$

6. Plot a straight line over the test data between $1\,\text{h}^{-1/2}$ and $2.24\,\text{h}^{-1/2}$ on the inverse square root of the time graph, Fig. 2.45.

Note: Time increases backward on the inverse square root of the time plot (i.e., from right to left). Determining the correct section of the test data on the inverse square root of the time plot is practically impossible without using the log-log diagnostic plot to determine the beginning and end of the spherical flow phase.

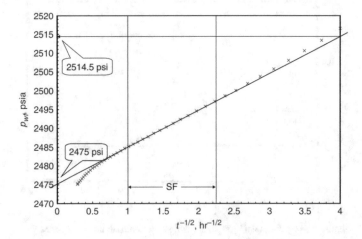

Fig. 2.45 Show the inverse square root of the time graph, showing spherical/radial flow

7. Calculate the slope m_S from any two points on the straight line during the spherical flow period:

$$|m_s| = \left| \frac{2514.5 - 2475}{4 - 0} \right| = 9.9\,\text{psi} \cdot \text{h}^{1/2}$$

8. Determine the spherical permeability:

$$k_s = \left(\frac{4906\ qB \times \sqrt{\emptyset \mu c_t}}{m_s} \right)^{2/3}$$

$$k_s = \left(\frac{4906 \times 625 \times 1.21 \times 1.06 \times \sqrt{0.27 \times 1.06 \times 12.9 \times 10^{-6}}}{9.9} \right)^{2/3} \mu = 84 \text{ mD}.$$

9. Determine the vertical permeability:

$$k_v = \frac{k_s^3}{k_h^2} = \frac{(84)^3}{(141)^2} = 29.4 \text{ mD}.$$

Note: The permeability determined from the radial flow is the horizontal permeability k_h.

10. Determine the radius of investigation at the start and end of spherical and radial flow periods.

$$r_{ibsf} = \sqrt{\frac{k_v t_{bsf}}{948 \emptyset \mu c_t}} = \sqrt{\frac{29.4 \times 0.2}{948 \times 0.27 \times 1.06 \times 13 \times 10^{-6}}} = 41 \text{ ft}$$

$$r_{iesf} = \sqrt{\frac{k_v t_{esf}}{948 \emptyset \mu c_t}} = \sqrt{\frac{29.4 \times 1}{948 \times 0.27 \times 1.06 \times 13 \times 10^{-6}}} = 92 \text{ ft}$$

$$r_{ibrf} = \sqrt{\frac{k t_{brf}}{948 \emptyset \mu c_t}} = \sqrt{\frac{141 \times 2}{948 \times 0.27 \times 1.06 \times 13 \times 10^{-6}}} = 284 \text{ ft}$$

$$r_{ierf} = \sqrt{\frac{k t_{erf}}{948 \emptyset \mu c_t}} = \sqrt{\frac{141 \times 12}{948 \times 0.27 \times 1.06 \times 13 \times 10^{-6}}} = 695 \text{ ft}$$

Note: At the beginning and ending of the spherical flow phase, the radius of investigation was calculated using the vertical permeability. Which enables us to analyze the distance between the distance of the perforations and the bottom of the reservoir thickness with the vertical change in pressure at the end of the spherical flow period. The well was completed at the top of the reservoir, so, the distance from the perforation center to the reservoir bottom is $dz = 100 - 15/2 = 92.5$ ft, which is consistent with the radius of investigation at the completion of the spherical flow period.

Also, the radius of investigation at the beginning and ending of the radial flow phase was determined using the horizontal permeability to provide the distance the transient changed in the horizontal direction at those points.

2.13 Bilinear Flow Regime

The combined simultaneous linear flow in vertical directions creates a bilinear flow regime. Such a flow regime is commonly found in hydraulically fractured well testing, and it occurs for finite-conductivity fractures with linear flow in the fracture and to the fracture plane. On the logarithmic derivative diagnostic plot, the bilinear flow phenomenon is represented by a 1/4 slope line. Its occurrence enables the assessment of fracture conductivity.

The bilinear flow regime can occur under the following conditions:

- A horizontal well having transitory dual porosity behaviour during the intermediate linear flow phase in a fractured or layered reservoir,
- A vertical well near a high conductivity fault, and
- A vertical well that is placed between two parallel leaky barriers caused by faulting or sedimentary processes.

As seen in Fig. 2.46, the log–log derivative curve for a bilinear flow has a slope of 1/4. If the fracture is undamaged, the pressure change curve will similarly have a slope of 1/4 but it will be four times higher than the derivative curve. The pressure change curve will roughly shift toward a 1/4 line from above if the fracture is damaged.

Through bilinear flow, the pressure response is a linear function of the fourth root of time, and can be expressed as:

$$\Delta p = m_B t^{1/4} + b_B \tag{2.67}$$

On a plot of pressure versus the fourth root of time, test results demonstrating bilinear flow will follow a line with a slope of m_B and an intercept of b_B. The slope m_B for a vertical fracture with finite conductivity is inversely propisitional to the square root of the fracture conductivity.

From the log–log derivative, the bilinear flow derivative may be found by using the equation below:

$$\frac{dp}{dt^{1/4}} = \frac{4}{t^{1/4}}\left(t\frac{dp}{dt}\right) \tag{2.68}$$

In the case of finite-conductivity vertical fracture, if the reservoir permeability is identified, the slope m_B can be utilized to determine the fracture conductivity. The slope m_B can be calculated from the straight-line section of a plot of pressure versus the fourth root of time. Also, the slope m_B can be determined from the field test data derivative of the log–log diagnostic graph using the following equation:

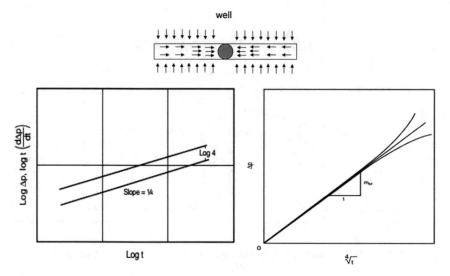

Fig. 2.46 Pressure and the derivative response of bilinear regime behaviour follow a line with a slope of 1/4 on the log–log plot

$$m_B = \frac{4}{t_B^{1/4}} \left(t \frac{dp}{dt} \right)_B \tag{2.69}$$

where $(tdp/dt)_B$ is the logarithmic derivative and t_B is the time, for any point on the slopw 1/4 line relating to the bilinear flow regime. Also, m_B can be read from the horizontal line of the derivative on the bilinear flow diagnostic graph.

The bilinear flow regime is most commonly found in hydraulically fractured wells, but it can also be found in other flow conditions, such as from a horizontal well in a transient dual-porosity reservoir, a vertical well in a channel reservoir with leaky boundaries, or a well near a finite-conductivity fault (Du and Stewart 1995).

The next sections, will discuss about the bilinear flow regime for a well with a finite-conductivity hydraulic fracture and a well located near a single, finite-conductivity fault.

2.13.1 Finite-Conductivity Hydraulic Fracture

This flow is visible during hydraulically fractured well testing with supported fractures, and it is a crucial aspect of typical well test analysis. The pressure response of a well with a finite-conductivity vertical fracture in a bilinear flow regime is shown in Fig. 2.47.

The pressure response for bilinear flow in a finite-conductivity fracture is presented in dimensionless form by Cinco-Ley and Samaniego (1981), as follows:

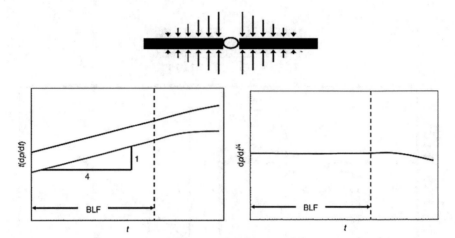

Fig. 2.47 Well with finite conductivity vertical hydraulic fracture, showing bilinear flow

$$p_D = \frac{\pi}{\Gamma(1.25)\sqrt{F_{cD}}}t_{LfD}^{1/4} + S_f = \frac{2.45}{\sqrt{F_{cD}}}t_{LfD}^{1/4} + S_f \qquad (2.70)$$

where F_{cD} is the dimensionless fracture conductivity which is defined as:

$$F_{cD} = \frac{w_f k_f}{k l_f} \qquad (2.71)$$

and t_{LfD} is defined as a dimensionless time based on the hydraulic fracture half-length, L_f, which is expressed as:

$$t_{LfD} = \frac{0.0002637\,kt}{\emptyset \mu c_t L_f^2} \qquad (2.72)$$

where k_f is the permeability of the proppant in the fracture, w_f is the fracture width, and L_f is the fracture length.

In oilfield units the pressure response can be written as:

$$\Delta p = \frac{44.1\,q B \mu}{h\left(w_f k_f\right)^{0.5}(\emptyset \mu c_{tk})^{0.25}}t^{0.25} + \frac{141.2\,q B \mu}{kh}S_f \qquad (2.73)$$

If the reservoir permeability is identified, test data in bilinear flow could be used to calculate the fracture conductivity from the bilinear flow slope m_B using the following equation:

$$w_f k_f = \left(\frac{44.1 q B \mu}{k|m_B|}\right)^2 \sqrt{\frac{1}{K \emptyset \mu c_t}} \qquad (2.74)$$

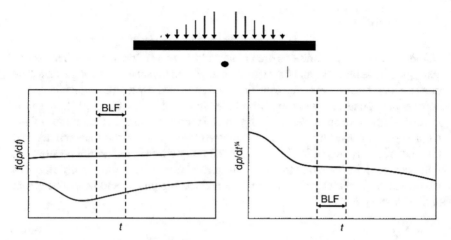

Fig. 2.48 Pressure and pressure derivative response, along with the bilinear flow diagnostic plot

2.13.2   *Well Near a Finite-Conductivity Fracture*

For a well close to a single, infinitely long fracture with finite conductivity, Fig. 2.48 shows the pressure and pressure derivative response as well as the bilinear flow diagnostic graph.

The model is comparable to that of a well with a low-conductivity vertical fracture with choked fracture skin s_f, which is determined by the following equation:

$$S_f = \ln \frac{2L}{r_w} \tag{2.75}$$

where L is the distance from the well location to the finite-conductivity fracture.

2.14   Other Flow Regimes

In addition to the well-known flow regimes described previously, there are other types of flow regime behaviour that can generate a logarithmic derivative with a power-law functional form. For wells approaching constant pressure boundaries, flow patterns which including radial stabilization (slope $= -1$), linear stabilization (slope $= -1/2$), and spherical stabilization (slope $= -3/2$) can occur. In addition, radial, linear, and spherical stability can result in an infinite radial, linear, or spherical system within a short flow time.

2.15   Summary

Numerous boundary conditions are connected to various flow regimes. The steady state, pseudosteady state, and transient state flow regimes are often distinguished. The rate at which pressure changes over time indicates the flow regime, which is dependent on the boundary condition. This chapter covered all the forms of flow regime in vertical and horizontal wells, at early and late redial/linear/spherical/bilinear flow, including flow caused by the hydraulic created fracture. Also, the chapter discussed high-conductivity hydraulic fracture, limited-entry completion, partial penetration, finite-conductivity hydraulic fracture, and channel reservoirs. Furthermore, exercises are provided in several sections to enable engineers to understand flow concepts in various flow systems.

References

Camacho-V., R.G., Raghavan, R., and Reynolds, A.C. 1987. Response of Wells Producing Layered Reservoirs: Unequal Fracture Length. SPE Form Eval 2 (1): 9–28. SPE-12844-PA. https://doi.org/10.2118/12844-PA

Chatas, A.T. 1966. Unsteady Spherical Flow in Petroleum Reservoirs. SPE J. 6 (2): 102–114. SPE-1305-PA. https://doi.org/10.2118/1305-PA

Cinco-Ley, H., and F. Samaniego-V., 1981, Transient pressure analysis for fractured wells: Journal of Petroleum Technology, v. 33, no. 9, p. 1749–1766. https://doi.org/10.2118/7490-PA

Cinco-Ley, H. and Samaniego-V., F. 1981. Transient Pressure Analysis: Finite Conductivity Fracture Versus Damaged Fracture Case. Presented at the SPE Annual Technical Conference and Exhibition, San Antonio, Texas, USA, 4–7 October. SPE-10179-MS. https://doi.org/10.2118/10179-MS

Cinco-Ley, H., Samaniego-V., F., and Dominguez-A., N. 1976. Unsteady-State Flow Behaviour for a Well Near a Natural Fracture. Presented at the SPE Annual Fall Technical Conference and Exhibition, New Orleans, 3–6 October. SPE-6019-MS. https://doi.org/10.2118/6019-MS

Du, K. and Stewart, G. 1995. Bilinear Flow Regime Occurring in Horizontal Wells and Other Geological Models. Presented at the International Meeting on Petroleum Engineering, Beijing, 14–17 November. SPE-29960-MS. https://doi.org/10.2118/29960-MS

Ehlig-Economides, C. and Economides, M.J. 1985. Pressure Transient Analysis in an Elongated Linear Flow System. SPE J. 25 (6): 839–847. SPE-12520-PA. https://doi.org/10.2118/12520-PA

Ehlig-Economides CA, Hegeman P, Vik S (1994) Guidelines simplify well test analysis. Oil Gas J 92:33–40.

Proett, M.A. and Chin, W.C. 1998. New Exact Spherical Flow Solution With Storage and Skin for Early-Time Interpretation With Applications to Wireline Formation and Early-Evaluation Drillstem Testing. Presented at the SPE Annual Technical Conference and Exhibition, New Orleans, 27–30 September. SPE-49140-MS. https://doi.org/10.2118/49140-MS

Spivey, John P. and W. John Lee. 2013. Applied Well Test Interpretation. SPE Textbook Series, volume 13.

Stewart, G. 2011. Well Test Design and Analysis. Tulsa: PennWell. Teng, D.T., Maloney, B.J., and Mantecon, J.C. 2006. Well Test by Design: Transient Modelling to Predicting Behaviour in Extreme Wells. Presented at the SPE Asia Pacific Oil & Gas Conference and Exhibition, Adelaide, Australia, 11–13 September 2006. SPE-101872-MS. https://doi.org/10.2118/101872-MS

Chapter 3
Well Test Interpretation Workflow

3.1 Introduction

Pressure transient testing is a crucial component of reservoir management and is one of the main assessment techniques. Reservoir monitoring and surveillance depend on obtaining a large amount of data, such as production profile, pressure data, fluid saturation distribution, along with reservoir core and fluid samples, etc. Such data are evaluated periodically and help the engineers to create good decision making.

The described workflow must be relevant to all pressure-transient testing conditions. As a result, it is uncommon for any test to require every step in the workflow. However, certain tests may need extra steps not included in this workflow. The processes described below are given in the order in which they would usually be performed; it is frequently required to repeat between steps to finish the interpretation. The workflow steps are as follows:

1. Gather the data required for the interpretation,
2. Analysis, quality control, and setting up the data for interpretation,
3. Break up the test data based on the change in the pressure responses,
4. Characterize the flow regimes that appear in the test data,
5. Choose the reservoir model to apply for interpretation,
6. Determine the parameters that describe the reservoir model by applying manual straight-line and log–log approaches,
7. Mimic or history-match the pressure response,
8. If proper, estimate confidence intervals,
9. Interpret the computed model parameters, and
10. Confirm the results.

© The Author(s), under exclusive license to Springer Nature Switzerland AG 2023   77
T. A-A. O. Ganat, *Modern Pressure Transient Analysis of Petroleum Reservoirs*,
Petroleum Engineering, https://doi.org/10.1007/978-3-031-28889-0_3

3.2   Well Testing Workflow Steps

3.2.1   Gather Well Test Data

The first thing required to interpret a pressure-transient test is to assemble the test data. The following are the required data for well test interpretation:

- Well data, such as type of well completion, drilling report, simulation type, type of artificial lift used, wellbore schematic, etc.
- Reservoir data, such as pressure, temperature, datum, fluid contacts, etc.
- Test data, such as test reports, pressure gradient data, and variation of fluid rate versus time.
- Detailed flow rate data during the test period.
- Gauge data, such as variation of pressure and temperature data versus time, start and end test date, gauge type, etc.
- Fluid Property data,
- Petrophysics data, such as Logs, pore volume compressibility, Core analysis data, etc.
- Geology and Geophysics such as structure maps, boundaries, natural fracturing, layering, fluid contacts, etc.

3.2.2   Analysis and Quality Control Data

The quality control and validating the obtained test data is the most important stage before progressing with the interpretation. Mattar and Santo (1992) suggest that over 50% of the evaluating time should be dedicated to investigating, validating, and settling the raw data. The following are the main data that need to be reviewed before establishing the well test interpretation:

- Evaluate the rate-time data, such as if the test data is complete or not, if there are any nearby wells that may affect the pressure response, if the fluid ratios are steady before the test, during the test, etc.
- Analysis Gauge Data. Apply a Cartesian scale to the gauge data evaluation. Be specific about the beginning and ending of each flow regime period. Compare the data if there are multiple gauges in the hole and select the more accurate gauge data.
- Determine pressure change and derivatives for every single test flow period.
- Pinpoint non-reservoir phenomena and the interpretation must in some way account for the phenomenon. There are three choices, either ignore it, model it, or remove it from the selected data.
- Find the functions for plotting. The pressure plotting tool has two main options: pressure and rate-normalized pressure change. Elapsed time, Agarwal multi-rate equivalent time, logarithmic superposition time function, and Horner time ratio

are some options for time-plotting functions for build-up testing. Elapsed time and material balance time are two methods that time can be plotted during flow testing.
• Search for data showing any depletion and identifies the effect of nearby wells.

3.2.3 Deconvolve Data

High awareness is needed for efficient use of deconvolution, which is uncertain procedure without a unique solution. For more information, see Levitan et al. (2006)s instructions for improving deconvolution performance. Use of build-up data for deconvolution, synchronization of rate and pressure data, removal of nonlinear phenomenon-related data distortions, use of the entire rate history, deconvolution of every build-up separately, and adjusting initial pressure for consistency between build-ups are some of these rules.

3.3 Indicate Flow Regimes

Locate any flow regimes displayed during the well test data. Normally, the log–log diagnostic graph is used for flow-regime identification. Additionally, the flow-regime-specific diagnostic plots may be used to identify the flow-regime. Only flow-regime identification as determined by the log–log diagnostic plot and the flow-regime specific diagnostic plots can be used as the basis for the use of the flow-regime specific straight-line plots. Table 3.1 show the summary of the flow regime's diagnostic characteristics.

Table 3.1 Summary of flow regime diagnostic characteristics

Flow regime	Slope of logarithmic Derivative on log–log plot	Flow-regime specific Diagnostic plot	Flow-regime specific Straight-line analysis plot
Radial	0	$dp/ln(dt)$ versus Δt	p versus log (Δt)
Linear	1/2	dp/dt versus Δt	p versus $\Delta t^{1/2}$
Volumetric	1	dp/dt versus Δt	p versus Δt
Spherical	$-1/2$	$dp/dt^{-1/2}$ versus Δt	p versus $\Delta t^{-1/2}$
Bilinear	1/4	$dp/dt^{1/4}$ versus Δt	p versus $\Delta t^{1/4}$

3.4   Select Reservoir Model

Effective well test interpretation depends on the use of a suitable reservoir model. Most often, inaccurate interpretations result from using the wrong reservoir model, inappropriate data analysis, or incorrect flow regime identification. Making these errors will result in poor reservoir management. The selection of a model is mostly driven by engineering data and outside data.

Making better reservoir management decisions largely depends on the quality and amount of data obtained for a well test interpretation. Kikani (2009) proposed a comprehensive analysis of the data obtained from well testing.

3.5   Estimate Model Parameters

Analyzing the parameters that define the reservoir model is quite simple if a reservoir model has been carefully selected. These three categories can include practically all parameter estimation approaches, including straight-line methods, log–log methods, simulation or history matching procedures, and others. Although any of these approaches may be used to gather the reservoir's attributes, the preferred methodology calls for applying straight-line or log–log methods to estimate the reservoir's parameters roughly before manually or automatically performing history matching to enhance those parameters.

Using a straight line has the following drawbacks:

- The corresponding flow regime must be appearing for each straight-line approach.
- For straight-line methods, only use the data from one flow regime and disregard the remaining data.
- For superposition time functions presume that only the corresponding flow regime occurs.

3.6   Simulate or History-Match Pressure Response

Simulate or history-match pressure response is a very essential process of the well testing interpretation workflow, that is always ignored during the evaluation process.

Figure 3.1 illustrates the history matching workflow in terms of a typical reservoir modeling process. The various reservoir models used in history matching, whether for a static or dynamic model, result in the best reservoir data matching. History matching is only one aspect of uncertainty quantification, which is taken into consideration as a whole. It is necessary to assess the result's uncertainty before making a decision. Most assessments of uncertainty are subjective. History matching should assist in developing the uncertainty model, which is then used in decision-making.

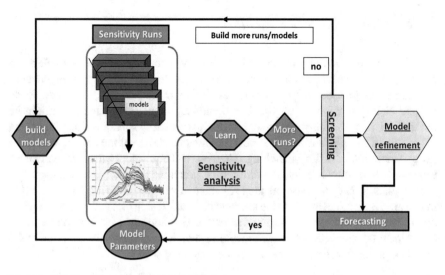

Fig. 3.1 Graphic of general modeling workflow

3.6.1 Simulation

The most common approach to account for varying rate histories and complicated reservoir models is to use analytical or numerical simulation together with manually or automatically matched historical data. Analytical simulation avoids the presumptions needed to use straight-line or type-curve procedures, such as the use of superposition time functions, pressure and time transforms, or pressure and time transforms. Many boundary conditions and nonlinear events that lack analytical solutions can be simulated using numerical simulation.

In analytical modeling, the pressure response for a given flow-rate history is calculated using the dimensionless solution to the diffusivity equation at a constant flow rate for a certain set of inner and outer boundary conditions and reservoir pores media. The analytical simulation is often more accurate for single-phase systems with small and constant compressibility. It is possible to apply analytical modeling for single-phase gas reservoirs under the appropriate conditions, even accounting for stress-dependent porosity and permeability as well as the desorption of gas from the matrix in naturally fractured shale reservoirs (Al-Hussainy et al. 1966). Numerical simulation may be used to successfully describe any reservoir flow system, including linear and nonlinear flow, multiphase flow, vertical and lateral heterogeneities, and all boundary geometries. However, utilizing numerical simulation, the collection of the data and the history matching take much more time (Spivey and Semmelbeck 1995).

3.6.2   Manual History Matching

When manually analyzing reservoir histories, the known reservoir parameters are normally kept constant while the unknown reservoir variables are changed until the calculated pressure response matches the observed pressure response. To do manual history matching, analytical or numerical simulation techniques might be utilized. History matching must match on all history using superpositions. Verifying the results of straight-line and type-curve approaches is one of the crucial reasons of using simulation or history matching. The Agarwal equivalent time and the Horner time ratio are examples of superposition time functions that should be avoided.

The phases that are recommended for manual history matching to find a rapid match are displayed below. Use the pressure change log–log plot and pressure derivative from stages 1 through 4 during the match. Use both the Log–Log plot and the Cartesian plot in stage 5. After each method's match is complete, fix the value of the modified parameter in that stage for all the following stages.

Step 1: Change the WBS coefficient to match the WBS data.
Step 2: Change the permeability to get the simulated pressure derivative to match the horizontal line of the field data derivative.
Step 3: During the period related with IARF, adjust the skin factor to ensure the simulated pressure change match to the pressure change in the field data that was observed.
Step 4: Modify the boundary distances until the simulated pressure derivative matches the field-data derivative.
Step 5: Using the Cartesian plot, adjust the initial pressure and reservoir volume for a perfect fit.
Step 6: To adjust the permeability and skin factor while keeping the productivity index constant, determine the new skin factor from the following equation:

$$S_{new} = (S_{old})\left(\frac{k_{new}}{k_{old}}\right) - 8 \qquad (3.1)$$

3.7   Validate Results

Validation of the interpretation results is another critical stage in the workflow. The validation procedure involves the following:

- Conducting simple checks to ensure that the findings produced are acceptable,
- Determine whether the calculated parameter from several flow times is steady,
- Comparing test findings obtained with earlier tests done on the same well,
- Calculate the radius of investigation at the start and end of each flow regime,
- Simulating the whole rate and pressure history and compare it to the observed pressure history, and
- Using external data to compare parameter estimations.

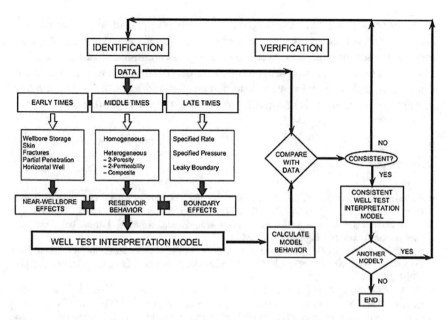

Fig. 3.2 Show the interpretation model identification procedure

If there are still disparities after re-analyzing various interpretations of external data, the proposed model must be ignored, and interpretation should be tried using an alternative model.

Typically, a single flow regime component that controls the flow period at various times is combined with other single flow regime components to produce the interpretation model.

Even if there are some potential components for an interpretation model, their combination might result in a wide range of interpretation models that correspond to the observed well behaviour. The difficulty for the well test interpreter is to determine which variables of the observed well performance should be included in the interpretation model. Figure 3.2 illustrates the entire interpretation process in a flowchart.

3.8 Summary

Since well and reservoir tests are the sources of essential data for reservoir models, engineers typically utilize them to validate or adjust the parameters of the models. By applying these models to better understand the interactions between reservoir fluids, formation, and well, engineers may be able to use this data to develop better completion and development strategies. Therefore, well test interpretation, which is the practice of understanding about a reservoir by assessing the pressure transient response brought on by a change in production rate, is crucial in determining how to manage a reservoir as a whole. This chapter thoroughly discussed the principles and stages of the test interpretation workflow. There was additional detail presented on how to collect data from well tests, estimate model parameters, analyze and control quality data, and deconvolve data. Also, the chapter addresses the reservoir model selection, simulation or history-match pressure response, and validation.

References

Al-Hussainy, R., Ramey, H.J. Jr., and Crawford, P.B. 1966. The Flow of Real Gases Through Porous Media. J. Pet Tech 18 (5): 624–636. SPE-1243-A-PA. https://doi.org/10.2118/1243-A-PA.
Kikani, J. 2009. Value of Information. In Transient Well Testing, ed. M.M. Kamal, Vol. 23, 53–64. Richardson, Texas: Monograph Series, SPE.
Levitan, M.M., Crawford, G.E., and Hardwick, A. 2006. Practical Considerations for Pressure-Rate Deconvolution of Well-Test Data. SPE J. 11 (1): 35–47. SPE-90680-PA. https://doi.org/10.2118/90680-PA.
Mattar, L., and Santo, M. 1992. How Wellbore Dynamics Affect Pressure Transient Analysis. J. Cdn. Pet. Tech. 31 (2) 32–40. PETSOC 92-02-03. https://doi.org/10.2118/92-02-03.
Spivey, J.P. and Semmelbeck, M.E. 1995. Forecasting Long-Term Gas Production of Dewatered Coal Seams and Fractured Gas Shales. Presented at the Low Permeability Reservoirs Symposium, Denver, 19–22 March 1995. SPE-29580-MS. https://doi.org/10.2118/29580-MS.

Chapter 4
Well Test Design Workflow

4.1 Introduction

Well testing is one of the important exploration practices required for reservoir monitoring and surveillance. Well testing is used to successfully describe a complex reservoir structure, decrease uncertainty in the reservoir static and dynamic model, and confirm reservoir connectivity. The selection of the optimum well test duration is necessary to obtain enough data to characterize the reservoir with a high degree of confidence. This will generate more accurate reservoir dynamic model calibration.

Well test design and interpretation will maximize the value of well tests by combining both geological and geophysical models with dynamic well test data. This will increase confidence in reservoir models, enhance production forecasting, characterize reservoir connectivity, and find sweet spots. Also, the well test results will assist to interpret reservoir characterization with confidence in fractured environments. This helps reservoir engineers to model fractures explicitly and adjust with pressure transient test results to define which fractures matter and which do not. The well test is applicable to use for conventional, unconventional, and hydraulically or naturally fractured formations with multiple wells and multiple zones.

During the well test design process, it is very important to optimize the planned well test design for describing the geological characteristics of the reservoir in question and providing other well the test options. Ambiguity in the geological model is included in the design process to confirm best data quality and provide analysis that accurately reflects the reservoir characteristics.

T. A-A. O. Ganat, *Modern Pressure Transient Analysis of Petroleum Reservoirs*, Petroleum Engineering, https://doi.org/10.1007/978-3-031-28889-0_4

4.2   Define Test Objectives

Reservoir and well pressure transient testing objectives can be largely identified (Kamal et al. 1995) as:

1. Acquire reservoir pressure and temperature,
2. Take a fluid sample,
3. Determine reservoir in-situ permeability,
4. Estimate skin factor for productivity and injectivity case,
5. Determine relative permeabilities,
6. Find fluid contacts,
7. Determine mobility of fluid,
8. Estimate reservoir pressure gradients,
9. Estimate a minimum reservoir volume,
10. Determine well productivity index,
11. Determine the radius of investigation,
12. Determine storativity,
13. Obtain reservoir parting pressure,
14. Determine fluid types for every single flow unit,
15. Estimate the net to gross ratio of reservoir thickness,
16. Determine reservoir continuity and hydraulic communication,
17. Evaluate reservoir heterogeneities,
18. Identify reservoir inhomogeneities, (faults, fractures, etc.),
19. Assess fracture length and conductivity,
20. Define reservoir type (fractured or layered),
21. Evaluate reservoir extent and boundary types (no flow, pressure support),
22. Assess completion and stimulation productivity,
23. To quantify damage or stimulation
24. Evaluate near-wellbore and hydraulic fracture clean-up.

The listed objectives above are very comprehensive, but it is not final list. Many of the above objectives may also be acquired from other sources, such as permeability can be estimated from open hole cores and logs. So, the objective of the reservoir and well testing is to obtain transient data sets to decrease uncertainty and improve the reservoir dynamic model.

4.3   Fundamental Well Test Design Scenarios

Measurements must be made during well and reservoir testing as fluids flow from the hydrocarbon reservoirs. Throughout the whole life of an oil and gas field, such tests are performed at different stages, including exploration, development, production, and injection. The tests are carried out to determine whether or not the newly discovered reservoir will produce hydrocarbons at a rate that is acceptable for commercial

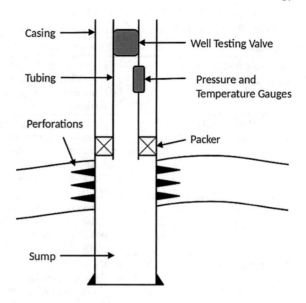

Fig. 4.1 A schematic of the siting of the downhole valve and gauge

use. Additionally, the test results will be utilized to define the reservoir's boundaries and identify the most cost-effective methods for producing wells and fields.

There are usually multiple well-tested scenarios. These scenarios usually differ in their objectives, the amount of data available, the cost of designing and performing a test, and the decisions to be taken based on the test results. Hence, the tests are conducted based on different circumstances, such as routine basis or to diagnose a specific problem. For instance, the most essential testing scenario is the drill-stem test (DST). For exploration wells, the drill-string is used as a temporary completion string to flow the well. In case the initial reservoir pressure and fluid type are unknown, a drill-string can be utilized in the case the well will be shut-in from downhole valve, as shown in Fig. 4.1, to prevent the reservoir fluid to flow to the surface (McAleese 2000). The quality of the obtained data will assist the engineers to make decisions about whether to abandon or develop the test layers or to perform an extended well test (Barnum and Vela 1984). Table 4.1 details the different scenarios to test the wells.

4.4 Alternative Well Testing

Normally, the objectives of well tests are conducted by using one of the scenarios stated in Table 4.1. However, there are other techniques/approaches to obtain the same data with equivalent quality at lower/same cost (Agarwal et al. 1999). Table 4.2 shows some of these alternative approaches.

Table 4.1 Below details the different scenarios to test the wells

Scenario	Objective
Drill-stem testing (DST) for exploration/appraisal well	• Collect the fluid sample • Determine reservoir pressure and temperature • Determine in-situ permeability • Formation evaluation
Extended well testing	• Determine hydrocarbons in place • Investigate reservoir boundaries • Estimate reservoir connectivity • Understanding reservoir drive mechanisms • Obtain more data that can't be acquired by short-term testing
Producing well	• Determine average drainage area pressure for material balance calculations, monitor movement of a fluid contact, diagnose specific productivity or injectivity problems
Development well	• Determine skin factors • Optimize completion strategies for subsequent wells • Collect permeability data for future use in reservoir modeling • Analyze reservoir connection in anticipation of implementing improved recovery
Well with rate-dependent skin factor	• To quantify rate-dependent skin factors by conducting a multi-rate test • Provides an inflow performance relationship • Estimates non-Darcy flow coefficient, permeability, and skin factor
Well with low productivity /Injectivity	• Determine the main reason for low or declining productivity or injectivity • Calculating skin factor • Effective permeability • Pressure in the average drainage area
Stimulated well	• Identify stimulation best practices for the reservoir • Determine reservoir permeability after stimulation

Table 4.2 Display alternative well testing approaches

Approach	Objective
Wireline formation tests	– Provide many reservoir permeability estimations – Provide reservoir pressure in relation to depth for around the same price as one conventional test
Production data analysis (Rate-Transient Analysis)	– Determine reservoir permeability – Estimating effective fracture half-length – Determine original fluid in place for low-permeability reservoirs

(continued)

Table 4.2 (continued)

Approach	Objective
Permanent downhole gauges	– Estimates of permeability – Estimate skin factor – Determine average reservoir pressure – Estimate the distances to fluid contacts
Log-derived permeability	– Estimates of permeability from logs A promising new method for predicting permeability from logs is presented by Anand et al. (2011), although the technique is not widely used

4.5 Collect Data

The main required data need to be collected for well test analysis is dominated fluid in the reservoir, PVT data, static data, flow rate data, and pressure data, as shown in Fig. 4.2.

The only data that is often available at the time of drilling exploratory wells is the exposed seismic data and the resulting static geological model. Open hole logs will be conducted in the exploratory well once it has been drilled. However, there won't be much time for in-depth interpretation before well testing begins. Chapter 3, Sect. 3.2 explains the collected data in more detail.

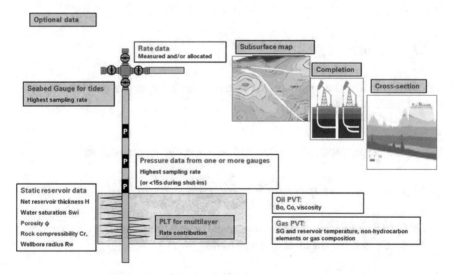

Fig. 4.2 Schematic diagram exhibiting the main information required for well test analysis

4.6   Estimate Reservoir Properties

Typically, the well test is conducted to obtain some of the reservoir properties. Thus, to test a particular well, the test design has to make some reasonable assumptions about the variables being investigated, such as the skin factor, wellbore-storage coefficient, boundary distances, and initial reservoir pressure.

4.7   Permeability Estimates

The reservoir is normally partitioned into regions with one well per region. The region is assumed to have its independent permeability. Clues to the volume over which permeability averaging occurs can be determined from the radius of investigation estimates and studies of composite reservoirs. Zones in which fluids do not flow cannot add to permeability estimates. So, the drainage radius, r_e, gives an outer bound to the area of the reservoir that can affect the permeability estimate. An estimate of the inner boundary of the zone of influence can be acquired from pressure transient testing in reservoirs with a circular discontinuity in permeability. A good well test design depends heavily on accurate permeability estimations, because permeability affects how long it takes to contact the reservoir boundaries, reach the end of the wellbore storage, and demonstrate that the minimum preferred fluid in place. Table 4.3 show different methods to estimate the reservoir permeability.

4.7.1   Example 4.1

Table 4.4 provides the required data to determine the reservoir permeability using the one-point method.

Solution
Determine the producing time (Horner pseudo-producing time estimate),

$$t_p = \frac{24\,G_p}{q_g} = \frac{24 \times 2093}{6278} = 8 \text{ hr}$$

Using trial and error method to estimate the permeability (use different iterations), First, calculate the transient drainage radius.

$$r_d = \sqrt{\frac{kt}{377 \emptyset \mu c_t}} = \sqrt{\frac{0.1 \times 8}{377 \times 0.15 \times 0.0209 \times 1.78 \times 10^{-4}}} = 62 \text{ ft.}$$

Table 4.3 Display the methods to estimate the reservoir permeability

Productivity index	In case the single-phase productivity index, J, is known, $k = \frac{141.2 J q B \mu}{h}\left[\ln\left(\frac{r_e}{r_w}\right) - \frac{3}{4} + S\right]$ (4.1)
Data in pseudo-steady-state flow	For pseudo-steady-state flow from a closed circular reservoir, $k = \frac{141.2 q B \mu}{h(p_i - p_{wf})}\left[\ln\left(\frac{r_e}{r_w}\right) - \frac{3}{4} + S\right]$ (4.2) Equation 12.6 can be applied to calculate the permeability for single-point flow test data where the reservoir has hit pseudo-steady-state flow
Data in infinite-acting radial flow	For a single point of flow test data (Lee et al. 1984) is a simple method for an infinite acting reservoir $k = \frac{141.2 q B \mu}{h(p_i - p_{wf})}\left[\ln\left(\frac{r_d}{r_w}\right) - \frac{3}{4} + S\right]$ (4.3) where r_d is the transient-drainage radius $r_d = \sqrt{\frac{kt}{377 \emptyset \mu c_t}}$ (4.4)
Limitations of the one-point method	Because the one-point technique implies infinite-acting radial flow, the findings will be inaccurate if the test is completed before the end of wellbore storage or after reservoir boundaries are reached One-point limitations are: • Presumes infinite-acting radial flow • Needs a reliable independent assessment of the skin factor

Table 4.4 Given data

G_p	2093 MSCF
h	23 ft
ϕ	15%
p_i	3450 psi
r_w	0.33 ft
μ_{gi}	0.0209 cp
C_t	1.78×10^{-4} psi^{-1}
S	0
B_{gi}	0.810 BBL/MSCF
P_{wf}	2387 psia
q_g	6278 MSCF/D

Now, use the transient drainage radius and estimate the permeability,

$$k = \frac{141.2 q_g B_{gi} \mu_{gi}}{h(p_i - p_{wf})}\left[\ln\left(\frac{r_d}{r_w}\right) - \frac{3}{4} + S\right]$$

$$k = \frac{141.2 \times 6278 \times 0.810 \times 0.0209}{23 \times (3450 - 2387)} \left[\ln\left(\frac{62}{0.33}\right) - \frac{3}{4} + 0 \right] = 2.8 \, \text{mD}.$$

Next, use the estimated permeability and calculate the new transient-drainage radius,

$$r_d = \sqrt{\frac{2.8 \times 8}{377 \times 0.15 \times 0.0209 \times 1.78 \times 10^{-4}}} = 323 \, \text{ft}.$$

Then, calculate an updated permeability estimate,

$$k = \frac{141.2 \times 6278 \times 0.810 \times 0.0209}{23 \times (3450 - 2387)} \left[\ln\left(\frac{323}{0.33}\right) - \frac{3}{4} + 0 \right] = 3.8 \, \text{mD}.$$

Now use the obtained permeability from the second trail and calculate the new r_d, and estimate the new permeability,

$$r_d = \sqrt{\frac{3.8 \times 8}{377 \times 0.15 \times 0.0209 \times 1.78 \times 10^{-4}}} = 379 \, \text{ft}.$$

$$k = \frac{141.2 \times 6278 \times 0.810 \times 0.0209}{23 \times (3450 - 2387)} \left[\ln\left(\frac{379}{0.33}\right) - \frac{3}{4} + 0 \right] = 3.86 \, \text{mD}.$$

The estimated permeability from the third trail is only 3% greater than that from the second trail. So, the estimated permeability in the last trial is accepted.

4.8   Estimate Test Period to Reach Expected Flow Regime

The selection of the well flow rate is the main variable during the choice of the optimum well test period duration. In typical nodal analysis, the well inflow performance of the reservoir is modeled at a semi-steady state for a particular average pressure for the well drainage area using the following equation:

$$q_s = J_{sss}\left(P - P_{wf}\right) \tag{4.5}$$

where J_{sss} may be expressed as:

$$J_{sss} = \frac{2\pi k h}{B\mu\left(\ln\frac{r_e}{r_w} - \frac{3}{4} + S\right)} \tag{4.6}$$

However, in the case of an exploration well test, it is not sure that the flow regime reaches the state of semi-steady state depletion. It is more possible that the well will flow in transient conditions during the test. In this case, the inflow reservoir performance is modeled by the following transient productivity index:

$$q_s = J_t\left(P - P_{wf}\right) \tag{4.7}$$

where J_t may be expressed as:

$$J_t = \frac{2\pi kh}{B\mu\left(\frac{1}{2}\ln\frac{4kt_p}{\gamma\emptyset\mu r_w^2} + S\right)} \tag{4.8}$$

where t_p is the time relating to the duration of the main flow period.

4.8.1 Time to Beginning of Middle Time Region

In some tests, the wellbore storage is particularly a problem when testing low-permeability reservoirs. In such cases, the duration of the build-up will require at least 50–60 h to get an acceptable period of middle time region (MTR) straight line. Although these estimates of the wellbore storage coefficient are only approximate. Therefore, the nodal analysis is a main part of the well test design procedure. This design relates to a surface shut-in where the entire pipe string volume contributes to the storage effect. While in the case of a downhole shut-in, only the gas is trapped below the testing valve.

Typically, the time to end of wellbore storage is determined from the following equation (Earlougher 1977):

$$t_{eWBS} = \frac{(200{,}000 + 12{,}000S)\mu}{kh}C \tag{4.9}$$

The time can be written in dimensionless form as:

$$\left(\frac{t_D}{C_D}\right)_{eWBS} = 60 + 3.5\,S \tag{4.10}$$

where S is the total skin factor
the equation can be writing also as:

$$t_{eWBS} = \frac{200\,C}{JB} \tag{4.11}$$

Equations 4.9, 4.10 and 4.11 are only applicable to wells with a fully pene-trating completion, continuous wellbore storage, a skin factor of zero or one, and no

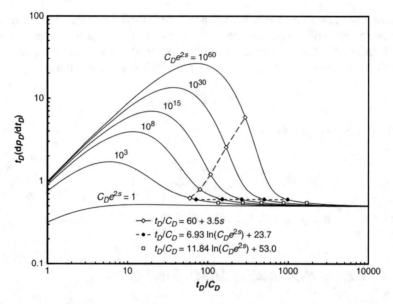

Fig. 4.3 Show constant wellbore storage duration and skin factor in an infinite-acting reservoir

non-Darcy skin. Before the logarithmic derivative was invented, Eqs. 4.9 and 4.10 were proposed. While Eq. 4.11 is appropriate for wells with minor or no damage, it understates the period until the end of wellbore storage when there is a significant mechanical skin factor, as seen in Fig. 4.3.

In the case of 100% non-Darcy skin to estimate the time to end of wellbore storage for the build-up test (see Fig. 4.4), Eq. 4.12 need to modify (for a 20% tolerance) as follows:

$$\left(\frac{t_D}{C_D}\right)_{eWBS} = 1.69(60 + 3.5S') \tag{4.12}$$

where S' is total skin factor, dimensionless.

In field unit:

$$t_{eWBS,ND} = \frac{(344,000 + 20,700S')\mu}{kh}C \tag{4.13}$$

In the case of 100% non-Darcy skin to estimate the time to end of wellbore storage for the drawdown test (see Fig. 4.4), Eq. 4.14 need to modify (for a 20% tolerance) as follows:

$$\left(\frac{t_D}{C_D}\right)_{eWBS} = 11\ln\left(C_De^{2S'}\right) + 41 \tag{4.14}$$

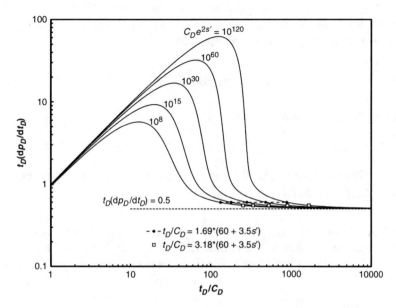

Fig. 4.4 Show the time to end of wellbore storage at 100% non-Darcy skin factor in an infinite-acting reservoir for a build-up test

In field unit:

$$t_{eWBS} = \left\{37{,}300 \ln\left(C_D e^{2S}\right) + 139{,}000\right\} \frac{\mu}{kh} C \qquad (4.15)$$

When transitioning from the wellbore storage period to infinite-acting radial flow, pressure response may show spherical or hemispherical flow if the well has a restricted entrance or partial penetration completion. The time at which spherical flow ends can be determined in oilfield units as:

$$t = 300 \frac{\emptyset \mu c_t h^2}{k_v} \qquad (4.16)$$

And for hemispherical flow time can be estimated from:

$$t = 1.200 \frac{\emptyset \mu c_t h^2}{k_v} \qquad (4.17)$$

where k_v = vertical permeability, mD.

If the well is stimulated, the time to the beginning of infinite-Acting radial flow can be estimated using the following equation:

$$t = 94,800 \frac{\emptyset \mu c_t r_{wa}^2}{k} \tag{4.18}$$

where r_{wa} is apparent wellbore radius, *ft*.

4.8.2   Time to Reach Particular Boundaries or Flow Regimes (Infinite-Acting Reservoir)

To achieve the test objectives for any flow regime, the test period must be chosen so that the characteristic derivative lasts at least 1/2 to 2/3 of the log cycle. Therefore, the test period should be at least three to five times the values provided by the following cases:

Well in a Closed Reservoir

$$t = 948 \frac{\emptyset \mu c_t r_e^2}{k} \tag{4.19}$$

Well in a Channel

$$t = 300 \frac{\emptyset \mu c_t w^2}{k} \tag{4.20}$$

Single No-Flow Boundary

$$t = 948 \frac{\emptyset \mu c_t L^2}{K} \tag{4.21}$$

where L is the distance to the boundary, ft, and w is the width of the channel, ft.

4.9   Estimation Test Period Based on Economical Evaluation

The entire test period should be adequate to ensure that an effective reservoir response has been acquired and proper interpretations can be achieved. But this may cause some practical and economical challenges for the flow period to be continued long enough for accurate interpretations. For instance, in build-up testing production loss can be a serious constraint or in drawdown testing the flow rate might not be able to maintain a constant during the test period.

Typically, in order to move on with the field development plan, the engineers must estimate the minimal productivity and reserves. Without knowing reservoir rock and fluid parameters other than average water saturation and total compressibility, it is possible to estimate the test duration required to demonstrate the minimal productivity index and initial oil I place (IOIP).

4.9.1 Estimation of the Minimum Commercial Productivity Index

The productivity index is a reasonable indicator to measure economic development, and this development can be restricted by the effects of variations in the geological characteristics of reserves to be developed.

The productivity index considers how productive potentialities are used: for instance, if there are scale effects, inefficiencies, or the degree of use capacity of these factors is appropriate.

To estimate the minimum productivity index, a combination of the below equations for the logarithmic derivative and productivity index during infinite-acting radial flow is applied:

$$\left(t\frac{dp}{dt}\right)_{IARF} = 141.2\frac{qB\mu}{kh}\left(t_D\frac{dp_D}{dt_D}\right)_{IARF} = 70.6\frac{qB\mu}{kh} \qquad (4.22)$$

If the required minimum permeability-thickness, $(kh)_{min}$, for a well to be produced economically, Eq. 4.22 can be rearranged as:

$$\left(t\frac{dp}{dt}\right)_{IARF} \leq 70.6\frac{qB\mu}{(kh)_{min}} \qquad (4.23)$$

The productivity index or infinite-acting radial flow can also be arranged based on the minimum economic productivity index, J_{min}, as follows:

$$(kh)_{min} = 141.2 J_{min} B\mu \left[\ln\left(\frac{r_e}{r_w}\right) - \frac{3}{4} + S'\right] \qquad (4.24)$$

Now combining Eqs. 4.23 and 4.24 as follows:

$$\left(t\frac{dp}{dt}\right)_{IARF} \leq \frac{q}{2J_{min}\left[\ln\left(\frac{r_e}{r_w}\right) - \frac{3}{4} + S'\right]} \qquad (4.25)$$

If there is no skin factor and assume typical well spacing, the final arrangement of Eq. 4.25 is:

$$\left(t\frac{dp}{dt}\right)_{IARF} \leq \frac{q}{16J_{min}} \tag{4.26}$$

4.9.2 Estimation of the Minimum Commercial Oil in Place

The initial oil in place equation is:

$$OOIP = \frac{Ah\emptyset\,(1-s_w)}{5.615\ B_O} \tag{4.27}$$

where the pore volume can be expressed as:

$$V_{P=Ah\emptyset} \tag{4.28}$$

By rearranging Eq. 4.27,

$$OOIP = \frac{V_P(1-s_w)}{5.615\ B_O} \tag{4.29}$$

In case the reservoir pore volume V_p is identified, the $OOIP$ can be estimated using Eq. 4.29.

The Cartesian derivative for the period of pseudo-steady-state flow (PSSF) can be written as:

$$\frac{dp}{dt} = \frac{0.234q\,B}{Ah\emptyset C_t} = \frac{0.234q\,B}{V_P C_t} \tag{4.30}$$

By combining Eqs. 4.29 and 4.30,

$$\frac{dp}{dt} = \frac{q(1-s_w)}{24\,OOIP\,C_t} \tag{4.31}$$

The Cartesian derivative to prove the minimum economical OOIP can be written as:

$$\left(\frac{dp}{dt}\right)_{PSSF} \leq \frac{q(1-s_w)}{24\,(OOIP)_{min}\ C_t} \tag{4.32}$$

In the case of logarithmic derivative, the equation can be written as:

$$\left(t\frac{dp}{dt}\right)_{PSSF} \leq \frac{q(1-s_w)}{24\,(OOIP)_{min}\,C_t}\,t \tag{4.33}$$

4.9.3 Test Period for Minimum Economics

The economical test period (for constant-rate, single-phase, and infinite-acting radial flow of a slightly compressible fluid) can be obtained by combining Eqs. 4.25 and 4.33 as follows:

$$\frac{q(1-s_w)}{24\,(OOIP)_{min}\,C_t}\,t = \frac{q}{2J_{min}\left[\ln\left(\frac{r_e}{r_w}\right) - \frac{3}{4} + S'\right]} \tag{4.34}$$

To rearrange Eq. 4.35, the minimum economical time can be expressed as:

$$t = \frac{24\,(OOIP)_{min}\,C_t}{2J_{min}\left[\ln\left(\frac{r_e}{r_w}\right) - \frac{3}{4} + S'\right](1-s_w)} \tag{4.35}$$

To simplify Eq. 4.35, assume the total skin factor is zero and standard well spacing, the test time in hours can be written as the following:
For oil,

$$t = \frac{3}{2}\frac{(OOIP)_{min}\,C_t}{J_{o,min}(1-s_w)} \tag{4.36}$$

For gas,

$$t = \frac{3}{2}\frac{(OGIP)_{min}\,C_t}{J_{g,min}(1-s_w)} \tag{4.37}$$

where OGIP is the original gas in place in MMSCF, $J_{o,min}$ is the productivity index for oil in STB/D/psi, and $J_{g,min}$ is the productivity index for gas in MMSCF/D/psi.

4.9.4 Estimation of the Minimum Economical Test Time Graphically

This method may also use to estimate the economical test duration time graphically. Equations 4.24 and 4.34 may be graphically shown to enable a quick evaluation of

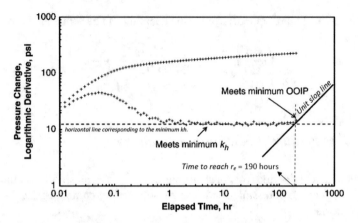

Fig. 4.5 Illustrates test meets both minimum *kh* and minimum *OOIP* conditions

the test period for constant-rate drawdown test data or deconvolved constant-rate pressure response data. On the standard diagnostic plot, Eq. 4.24 shows a horizontal line that provides the highest suitable value of the logarithmic derivative for the period of infinite-acting radial flow. The smallest economic OOIP is identified by a unit slope line in Eq. 4.34. Typically, the minimal OOIP is proven when the logarithmic derivative crosses the unit slope line.

Before conducting the test, the radius of the investigation needs to be estimated using Eq. 4.20. The equation gave a time of 190 h; therefore, the test duration was designed to be 220 h to reach the target. Figure 4.5 depicts an outstanding example of test conditions for a closed circular reservoir with the least economic OOIP and permeability-thickness product.

4.10   Estimate Test Rate and Determine Flow Rate Sequence

A prediction of the estimated flow rates, and a timing for the drawdown and build-up phases, should be included in the test design. To meet the test objectives, the flow rate or flow rates must be high enough to cause a properly measured pressure response. At the same time, the maximum flow rate must be minimal enough to match the rate achieved during the test. Typically, the following operational considerations restrict the flow rate during the test:

- *Flow rate and pressure constraints.* Some rate restrictions have an impact on the flow rate itself. The rate can also be regulated to minimize the overall pressure drawdown imposed on the reservoir or to manage the pressure at the wellhead or sand face. There is another limitation restricting the flow rate, for instance, capacity of the facilities, bubble point/ dew point pressure, minimum rate to lift

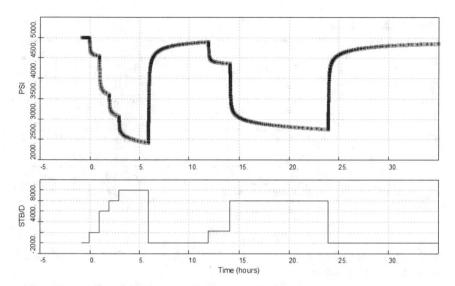

Fig. 4.6 Multi-rate test with the extended flow and build-up periods

the fluid, allowable disposal of produced fluids, sand production, and formation damage.

- *Maximum Sustainable Rate.* When determined the test period and maximum designed pressure drop, calculate the maximum flow rate that may be maintained during the test using the following equation:

$$q_{max} = \frac{k_{min}h}{162.6 \; B\mu\left(\log\left(\frac{k_{min}t}{\emptyset\mu C_t r_w^2}\right) - 3.23 + 0.869 \; S'\right)} \tag{4.38}$$

- *Flow Rate Sequence.* The flow rate sequence must be chosen to maximize the probability of meeting the test objectives. As illustrated in Fig. 4.6, the test design must include one major drawdown test period followed by a build-up test period if there is any purpose to conduct several flow rate tests and shut-in periods.

4.11   Example 4.2

A highly profitable gas reserve was found by a new exploratory well. According to the logs, the target zone depth ranges from 8195 to 8445 feet (TVD). With a temporary completion string that is just 50 feet open to flow in the pay interval, the well will be tested. Table 4.5 provides information on rock and fluid properties. The 8320 feet is the TVD (datum depth). According to the log data, the permeability should fall between 100 and 1000 mD. 10% of the original reservoir pressure should be the

Table 4.5 Show rock and fluid property data

h	250 ft
ϕ	27%
S (limited entry)	10
r_w	0.50 ft
B_g	0.810 bb/MSCF
μ_g	0.0225 cP
C_t	171×10^{-6} psi^{-1}
C_g	216×10^{-6} psi^{-1}
P_i	3,868.8 psi
Pressure gradient	0.465 psi/ft
Max allowed gas to flare	25 MMscf

maximum pressure drawdown allowed. A total test time of 72 hours is permitted under the well building budget.

Calculate the time to the end of wellbore storage using the following equation:

$$t_{eWBS} = \{40,000 \ln(C_D e^{2S}) + 180,000\} \frac{\mu}{kh} C$$

Assume $C_D e^{2S} = 2.25 \times 10^{62}$ and wellbore storage C = 0.065 bbl/psi.

$$t_{eWBS} = \{40,000 \times \ln(2.25 \times 10^{62}) + 180,000\}$$
$$\times \frac{0.0225}{100 \times 250} \times 0.065 = 0.346 \text{ hours}$$

Calculate the time to the end of spherical flow:

$$t_{eSF} = 300 \frac{\emptyset \mu c_t w^2}{k}$$
$$t_{eSF} = 300 \times \frac{0.27 \times 0.0225 \times 171 \times 10^{-6} \times (250)^2}{0.1 \times 100} = 1.95 \text{ hour}$$

Calculate the time to the beginning of MTR:

To estimate the test period, the minimum permeability estimates of 100 mD were used to determine the time to reach a radius of investigation of 5000 ft.

$$t = 948 \frac{\emptyset \mu c_t r_e^2}{k}$$
$$t = 948 \times \frac{0.27 \times 0.0225 \times 171 \times 10^{-6} \times (5.000)^2}{100} = 246 \text{ hour}$$

Calculate the radius of investigation for permeability of 100 mD, at the end of the 24-h flow period:

$$r_i = \sqrt{\frac{kt}{948\emptyset\mu C_t}}$$

$$r_i = \sqrt{\frac{100 \times 24}{948 \times 0.27 \times 0.0225 \times 171 \times 10^{-6}}} = 1.560\,\text{ft}$$

4.12 Summary

A workflow method is utilized during test design process to optimize test design for identifying geological characteristics of interest and presenting alternative test solutions. Uncertainty in the geological model is incorporated into the design process to provide optimal data quality and analysis as best represents the reservoir. This chapter will offer a detailed overview of the pressure transient test design process for well testing to maximize test time and decrease well switch-off period in a pilot test. Another objective is to validate the economic efficiency of the proposed process. The chapter thoroughly described the technique for designing well test scenarios, as well as well test objectives and data collection. Estimating reservoir characteristics and testing time based on economic assessment were discussed in detail with solved examples to help engineers understand the design process.

References

Agarwal, R.G., Gardner, D.C., Kleinsteiber, S.W. et al. 1999. Analyzing Well Production Data Using CombinedType-Curve and Decline-Curve Analysis Concepts. SPE Res Eval & Eng 2 (5): 478–486. SPE-57916-PA. https://doi.org/10.2118/57916-PA.

Anand, V., Freedman, R., Crary, S. et al. 2011. Predicting Effective Permeability to Oil in Sandstone and Carbonate Reservoirs From Well-Logging Data. SPE Res Eval & Eng 14 (6): 750–762. SPE-134011-PA. https://doi.org/10.2118/134011-PA.

Barnum, R.S. and Vela, S. 1984. Testing Exploration Wells by Objectives. Presented at the SPE Annual Technical Conference and Exhibition, Houston, 16–19 September 1984. SPE-13184-MS. https://doi.org/10.2118/13184-MS.

Earlougher, R.C. Jr. 1977. Advances in Well Test Analysis, Vol. 5. Richardson, Texas: Monograph Series, SPE.

Kamal, M. M., Fryder, D. G. and Murray, M. A. Use of Transient Testing in Reservoir Management. SPE-28002-PA, 1995.

Lee, W.J., Kuo, T.B., Holditch, S.A. et al. 1984. Estimating Formation Permeability From Single-Point Flow Data. Presented at the SPE Unconventional Gas Recovery Symposium, Pittsburgh, Pennsylvania, 13–15 May 1984. SPE-12847-MS. https://doi.org/10.2118/12847-MS.

McAleese, S. 2000. Operational Aspects of Oil and Gas Well Testing. Amsterdam, The Netherlands: Elsevier Science.

Chapter 5
Types of Well Tests

5.1 Overview

Well test selection and design present important concepts for designing oil and gas well tests, both pressure transient and well deliverability tests. Included in these design concepts are recommendations for selecting the appropriate well test to achieve the desired test objectives, estimating pre-test reservoir properties, selecting the proper flow-rate sequence for the test, and selecting the test period required to sample a chosen reservoir volume and or to reach stabilized flow conditions. Typical information obtained from well tests are permeability, distance to boundaries, size and shape of sand bodies, skin factor, and length of induced fractures.

Drawdown, build-up, and interference tests are the three main well tests. Additionally, there are tests for injection and falloff, which are comparable to the drawdown and build-up tests for injectors. In exploratory wells and newly drilled wells, a unique drawdown test called the Drill Stem Test (DST) is usually conducted.

5.2 Pressure Drawdown Test

The pressure Drawdown test (DDT) is just a sequence of bottom-hole pressure observations because of the constant producing rate. Usually, before the constant flowing rate, the well is closed for enough time to make the pressure widespread in the reservoir reach static reservoir pressure. The diagram showing this phenomenon is in Fig. 5.1.

The reservoir is initially kept at a constant pressure during the drawdown tests, and the well is originally closed off. The well is starting to flow, and the variations in the pressure response are being recorded over time. Normally, it is difficult to maintain a flow rate that is absolutely constant, and even little variations in flow rate during the test may significantly change the pressure response's trend. Rate normalization

Fig. 5.1 Schematic
illustration of pressure
drawdown

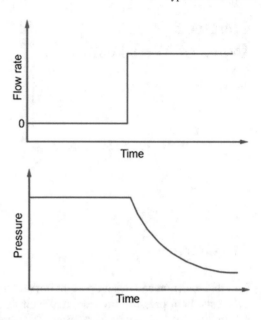

Fig. 5.1 Schematic
illustration of pressure
drawdown

is a technique that may be used to account for changing flow rates when the rate is gradually changing and the reservoir is infinite-acting at all relevant times.

5.2.1 Constant-Rate Production

The reservoir is at constant pressure throughout a constant-rate drawdown test since the well is shut off. The flowing bottom hole pressure (p_{wf}), which is measured as a function of time as the reservoir pressure drops, is generated by the well at a constant flow rate (q). As an infinite-acting reservoir, the equation for pressure response at a constant rate may be written as:

$$p_{wf}(t) = p_i - \frac{162.6q\,B\mu}{kh}\left[\log\left(\frac{kt}{\emptyset\mu C_t r_w^2}\right) - 3.23 + 0.869S\right] \quad (5.1)$$

The relation of $p_{wf}(t)$ versus log (t) will yield a linear trend with the downward direction line (Fig. 5.2) with slope m which can be expressed as:

$$m = -\frac{162.6q\,B\mu}{kh} \quad (5.2)$$

The intercept of the Y-axis, bottom-hole pressure, corresponds when ln$t$ equals zero when $t = 1$ h (Fig. 5.2). The corresponding pressure value is often written as $P_{t=1}$. Therefore, considering these procedures, the equation is simplified as:

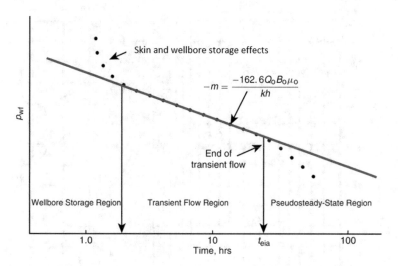

Fig. 5.2 Drawdown semi-log plot

$$y = p_{t=1h} = mx + b \qquad (5.3)$$

The intercept of this line with the Y-axis related to the value of pressure when t equals to 1 h is:

$$p_{t=1h} = p_i - \frac{162.6q\,B\mu}{kh}\left[\log\left(\frac{k}{\emptyset\mu C_t r_w^2}\right) - 3.23 + 0.869S\right] \qquad (5.4)$$

To adequately estimate skin factor, Eq. (5.1) can be re-written as follows:

$$S = 1.151\left\{\frac{p_i - p_{1h}}{\lfloor m \rfloor} - \log\left(\frac{k}{\emptyset\mu C_t r_w^2}\right) + 3.23\right\} \qquad (5.5)$$

So, from the semi-log plot, the permeability can be obtained from the slope m of the straight line and the skin factor can be determined to form the intercept at $p_{1\,h}$.

The additional pressure drop due to the skin effects can be expressed as:

$$\Delta p_{Skin} = 141.2\left(\frac{q\,B\mu}{kh}\right)S = 141.2|m|S \qquad (5.6)$$

5.2.1.1 Example 5.1

Estimate the oil permeability, skin factor, and the additional pressure drop due to the skin from the drawdown data of Fig. 5.3. Table 5.1 provides the reservoir data.

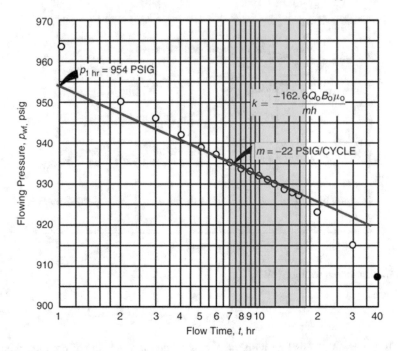

Fig. 5.3 Semi-log plot of pressure drawdown data

Table 5.1 Given data

h	130 ft
ϕ	20%
p_i	1154 psi
r_w	0.25 ft
μ_O	3.93 cp
C_t	8.74×10^{-6} psi^{-1}
B_o	1.14 BBL/STB
Q_o	348 STB/D

Assuming that the wellbore storage effect is not significant.

Solution

1. From Fig. 5.3, determine $p_{1\,h}$:

   $p_{1\,h} = 954$ psi

2. Determine the slope of the transient flow line:

   $m = -22$ psi/cycle

3. Determine the permeability by applying the following equation:

$$k = -\frac{162.6q\,B\mu}{mh}$$

$$k = -\frac{162.6 \times 348 \times 1.14 \times 3.93}{-22 \times 130} = 89\,\text{mD}$$

4. Estimate the skin factor by using the following equation:

$$S = 1.151\left\{\frac{p_i - p_{1h}}{m} - \log\left(\frac{k}{\emptyset\mu C_t r_w^2}\right) + 3.23\right\}$$

$$S = 1.151\left\{\frac{(1154 - 954)}{22} - \log\left(\frac{89}{0.2 \times 3.93 \times 8.74 \times 10^{-6} \times (0.25)^2}\right) + 3.23\right\}$$
$$= +4.6$$

5. Calculate the additional pressure drop:

$$\Delta p_{Skin} = 0.87 \times |m| \times S = 0.87 \times 22 \times 4.6 = 88\,\text{psi}$$

5.2.1.2 Example 5.2

Estimate the oil permeability, skin factor, and the additional pressure drop due to the skin from the drawdown data of Fig. 5.4. Table 5.2 provides the reservoir data.
Assuming that the wellbore storage effect is not significant.

Solution

1. From Fig. 5.5, determine p_{1h}:

$p_{1h} = 2060$ psi.

2. Determine the slope of the transient flow line:

$$m = \frac{2250 - 1930}{\log(0.001) - \log(100)} = -64\,\text{psi/cycle}$$

3. Determine the permeability by applying the following equation:

$$k = -\frac{162.6q\,B\mu}{mh}$$

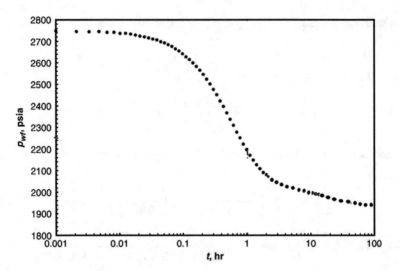

Fig. 5.4 Drawdown semi-log analysis example

Table 5.2 Given data

h	32 ft
ϕ	22%
p_i	2,750 psi
r_w	0.25 ft
μ_O	2.122 cp
C_t	10.9×10^{-6} psi^{-1}
B_o	1.152 BBL/STB
Q_o	125 STB/D

$$k = -\frac{162.6 \times 125 \times 1.152 \times 2.122}{-64 \times 32} = 24.3 \, \text{mD}$$

4. Estimate the skin factor by using the following equation:

$$S = 1.151 \left\{ \frac{p_i - p_{1h}}{m} - \log\left(\frac{k}{\emptyset \mu C_t r_w^2}\right) + 3.23 \right\}$$

$$S = 1.151 \left\{ \frac{(2750 - 2060)}{64} - \log\left(\frac{24.3}{0.22 \times 2.122 \times 10.9 \times 10^{-6} \times (0.25)^2}\right) + 3.23 \right\}$$

$$= +7.1$$

5. Calculate the additional pressure drop:

$$\Delta p_{Skin} = 0.87 x |m| \times S = 0.87 \times 64 \times 7.1 = 395 \, \text{psi}$$

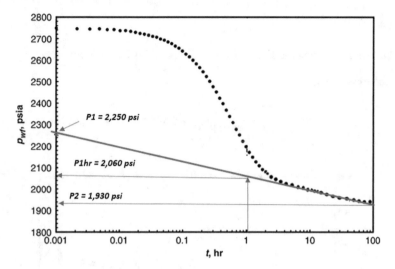

Fig. 5.5 Drawdown semi-log analysis

5.2.2 Smoothly Changing Rate

The rate-normalized pressure change can be used to assess data for a flow test when the rate is changing gradually and steadily. This is the condition when the rate is altering during the test. The evaluation of a variable flow rate drawdown test in an infinite-acting reservoir may be performed using the following equations:

The pressure response in an infinite-acting reservoir can be written as:

$$\frac{p_i - p_{wf}}{q(t)} = \frac{162.6B\mu}{kh}\left[\log\left(\frac{kt}{\emptyset\mu C_t r_w^2}\right) - 3.23 + 0.869S\right] \qquad (5.7)$$

The permeability k can be determined from the slope m using the following equation:

$$m = \frac{162.6B\mu}{kh} \qquad (5.8)$$

The skin factor equation can be written as follows:

$$S = 1.151\left\{\frac{\left(\frac{\Delta p}{q}\right)_{1h}}{m} - \log\left(\frac{k}{\emptyset\mu C_t r_w^2}\right) + 3.23\right\} \qquad (5.9)$$

5.2.2.1 Example 5.3

Evaluate the test shown in Fig. 5.6. Table 5.3 provides the reservoir data (Fig. 5.7).

1. Calculate the slope m.

$$m = \frac{78 - 51.5}{\log(10 - \log(0.01)} = 8.83\,\text{psi}/(\text{STB/D})/cycle$$

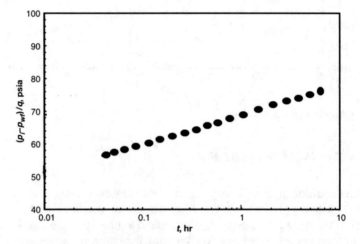

Fig. 5.6 Drawdown test, smoothly changing flow rate, with rate normalization

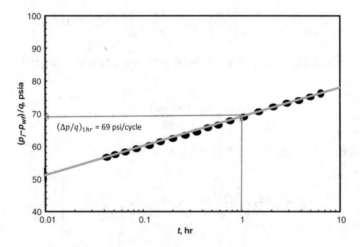

Fig. 5.7 Drawdown test, smoothly changing flow rate example

Table 5.3 Given data

h	35 ft
ϕ	18.5%
p_i	2750 psi
r_w	0.25 ft
μ_O	6.5 cP
C_t	8.42×10^{-6} psi^{-1}
B_o	1.152 BBL/STB

2. Estimate the permeability.

$$k = \frac{162.6 \times 1.15 \times 6.5}{8.83 \times 35} = 4\,\text{mD}$$

3. Calculate the skin factor.

$$S = 1.151 \left\{ \frac{69}{8.83} - \log\left(\frac{4}{0.185 \times 6.5 \times 8.42 \times 10^{-6} \times (0.25)^2} \right) + 3.23 \right\}$$
$$= +4.3$$

5.3 Type Curve Analysis

The type curve in the well test provides a method for evaluating the pressure draw-down and build-up tests using a graphic representation (Fig. 5.8). It is developed using a specific set of initial and boundary conditions and the analytical solution to the diffusivity equation. Dimensionless variables are used to provide the solutions displayed in the type-Curve. Dimensionless variables are used to describe the plot of these solutions.

There are various approaches that may be used to analyze any test for a vertical well of an infinite-acting homogeneous reservoir, such as the McKinley (1971) type curve, Earlougher and Kersch (1977) type curves, and Gringarten type curves (Gringarten et al. 1979).

According to Gringarten (1987), the most effective way to demonstrate the concept of a type curve is theoretical receipt during a test of any model interpretation that reflects the tested well and reservoir in visual analysis. Type curves are produced using the solutions to the flow equations in specific reservoir conditions. Type curves are shown as plots of dimensionless variables, such as dimensionless pressure vs dimensionless time.

For a very long time, the Gringarten type curve has been widely used (Gringarten et al. 1979). The type curves that are applied are based on a few presumptions,

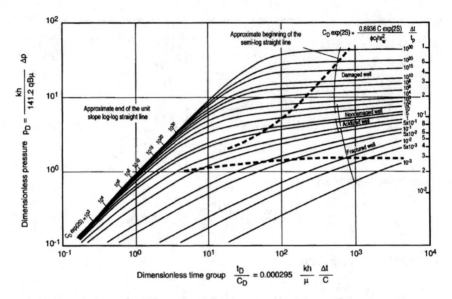

Fig. 5.8 Type curves for a well with wellbore storage and skin actor (homogeneous reservoir with infinite-acting)

including the constant vertical well production rate, single-phase flow with slightly compressible liquid flow, and homogenous reservoir with infinite-acting properties. Often, drawdown tests in undersaturated oil reservoirs make excellent use of the Gringarten type curve. Dimensionless variables, such as dimensionless pressure (p_D), dimensionless time (t_D/C_D), skin factor ($C_D e^{2S}$), and wellbore storage coefficient (C), represent the majority of the variables used. A log–log plot is used to illustrate the type curve with dimensionless pressure against time and wellbore storage coefficient parameters.

The pressure derivative was introduced by Bourdet et al. (1984), and the derivative type curve gives the graphical solution of the diffusivity equation. A log–log plot shows the derivative as a straight line.

Finding and matching the actual reservoir pressure response on the specific type curve is what is meant by "analyzing" the type-curve. As a basis of the graphical methods, the plot of real test data will be superimposed alongside the type curve, and the best fit is found depending on the type curve used. As a result, the dimensionless parameters that characterize the particular type curve may be used to predict reservoir and well parameters such as permeability and skin. In studies, the type-curve analysis has commonly been used.

5.3.1 Type-Curve Procedures Using Gringarten Type Curves

1. Plot the measured pressure drop Δp versus Δt on transparent paper lying on the type curves, using the log-log scale of the type curves.
2. Look for the part of underlying type curves that best matching of the data.
3. Note the specifications of the type curve where the measured points match; they relate to the value of $(C_D e^{2S})$.
4. Pick a match point, "M", whose coordinates can be read in both the type curve system of axes $(p_D, t_D/_D)$ and the field data system $(\Delta p, \Delta t)$. The point "M" can be picked anywhere on the plot, not necessarily on the curve. See the steps if Figs. 5.9, 5.10 and 5.11.

The ordinate of the match point is measured:

- In the type curve system of axes: p_D
- In the field data system of axes: Δp.

As:

$$p_D = \frac{kh}{141.2qB\mu}\Delta p \qquad (5.10)$$

The proportionality factor between p_D and Δp can be applied to estimate the reservoir's kh using the following equation:

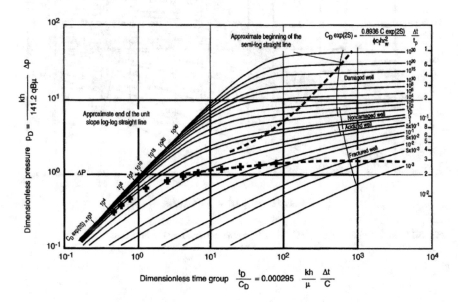

Fig. 5.9 Measured pressure drop Δp versus Δt on tracing paper lying on the type curves

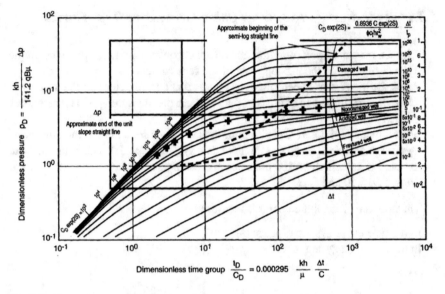

Fig. 5.10 Type curves matching the best data

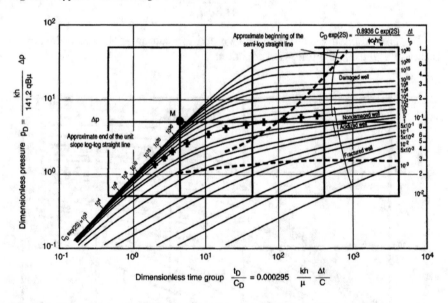

Fig. 5.11 Pick a match point, "*M*", after type curves matching the best data

$$kh = 141.2q\,B\mu \frac{(p_D)_M}{(\Delta p)_M} \tag{5.11}$$

In the same way the abscissa of the match point, "M", is measured in the type-curve system of axes, t_D/C_D and the field data system of axes, Δt. As kh is already obtained from Eq. 5.11, $\frac{t_D}{C_D}$ can be determined from the following equation:

$$\frac{t_D}{C_D} = \frac{0.000295kh}{\mu C}\Delta t \tag{5.12}$$

The proportionality factor between $\frac{t_D}{C_D}$ and Δt can be used to determine the wellbore storage, C, using the following equation:

$$C = \frac{0.000295kh}{\mu}\frac{(\Delta t)_M}{\left(\frac{t_D}{C_D}\right)_M} \tag{5.13}$$

The type curve where the data have been matched is characterized by $C_D\,e^{2S}$. The C_D can be estimated using the following equation:

$$C_D = \frac{0.8936C}{h\emptyset c_t r_w^2} \tag{5.14}$$

The value of $C_D\,e^{2S}$ is used to determine the skin factor using the following equation:

$$S = \frac{1}{2}\ln\frac{\left(C_D e^{2S}\right)}{C_D} \tag{5.15}$$

5.3.2 Example 5.4

An oil well was produced in a reservoir above bubble point pressure at a constant rate of 185 BOPD before it was shut-in for a build-up test. Table 5.4 provides the reservoir data.

Assume step 1: calculate Agarwal equivalent shut-in time, Δt_e, and pressure change, $\Delta p = (p_{ws} - p_{wf}$ @ $\Delta t = 0)$ and Step 2: Calculate pressure derivative concerning to the natural logarithm of equivalent time are already calculated and plotted in Fig. 5.12.

Table 5.4 Given data

h	114 ft
ϕ	28%
$P_{wf} \; at \; \Delta t$	2820 psi
r_w	0.5 ft
μ_O	2.2 cp
C_t	$4.1 \times 10^{-6} \; psi^{-1}$
B_o	1.1 BBL/STB
t_p	540 h

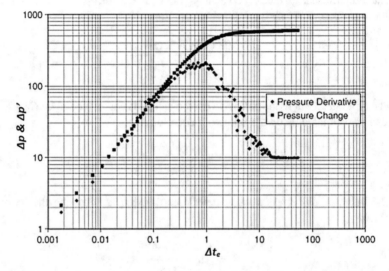

Fig. 5.12 Pressure and pressure derivative concerning to the natural logarithm of equivalent time

Calculate the formation permeability, k, the skin factor, S, the dimensionless well-bore storage, C_D, and the wellbore storage coefficient, C. Apply the type curve matching method using the combined Gringarten-Bourdet type curve.

Solution

Step 3: Determine C_D and C from the unit-slope line. Pick any point on the unit-slope line from Fig. 5.12. One such point on the unit-slope line is $(\Delta t_e \, /\Delta p)_{USL} = (0.06374/41.13)$.

$$C_D = \frac{0.03723 q \, B}{\emptyset C_t h_w^2} \left(\frac{\Delta t_e}{\Delta p} \right)_{USL}$$

$$C_D = \frac{0.03723 \times 185 \times 1.1}{0.28 \times 4.1 \times 10^{-6} \times 114 \times (0.5)^2} \times \left(\frac{0.06374}{41.13} \right) = 358.86$$

$$C = 0.04165q B \left(\frac{\Delta t_e}{\Delta p}\right)_{USL}$$

$$C = 0.04165 \times 185 \times 1.1 \times \left(\frac{0.06374}{41.13}\right) 0.013 \, RB/\text{psi}$$

Step 4: Perform type curve matching using the Figure and Gringarten-Bourdet type curve.

$$\left(\frac{p_D}{\Delta p}\right)_{PMP} = \left(\frac{4.5}{100}\right)$$

$$k = \frac{141.2q B \mu}{h} \left(\frac{p_D}{\Delta p}\right)_{PMP}$$

$$k = \frac{141.2 \times 185 \times 1.1 \times 2.2}{114} \times \left(\frac{4.5}{100}\right) = 25 \, \text{mD}$$

Step 5: Calculate C_D from the time match point (TMP).

$$\left(\frac{\Delta t_e}{\frac{t_D}{C_D}}\right)_{TMP} = \left(\frac{0.03}{1}\right)$$

$$C_D = \frac{0.03723k}{\emptyset \mu C_t r_w^2} \left(\frac{\Delta t_e}{\frac{t_D}{C_D}}\right)_{TMP}$$

$$C_D = \frac{0.03723k}{0.28 \times 2.2 \times 4.1 \times 10^{-6} \times (0.5)^2} \times \left(\frac{0.03}{1}\right) = 313$$

Step 6: Estimate skin factor, S.

$$S = 0.5 \ln\left(\frac{C_D e^{2S}}{C_D}\right)$$

where $C_D e^{2S} = 10^{21}$ from Fig. 5.13

$$S = 0.5 \times \ln\left(\frac{10^{21}}{313}\right) = 22.5$$

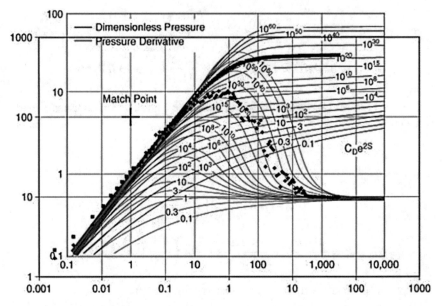

Fig. 5.13 Perform type-curve matching using Fig. 5.12 and Gringarten-Bourdet type curve

5.4 Pressure Buildup Test

The buildup test is the method of transient well testing that is most often performed. Closing the well and measuring the bottom-hole pressure are two steps in the analysis of pressure buildup. However, before closing the well, a steady flow rate must be achieved. This can be obtained from the well either at the beginning of production or during a time of continuous production to provide a steady pressure variation (Streltsova and McKinle 1984). In case the objective also determines the skin factor, then recording the pressure measurement before the shut-in of the well becomes necessary. Figure 5.14 show the well test in the appraisal well starting with a well clean-up period followed by two pressure BU tests and a main flow period.

Figure 5.14 show the well test is performed when the well has a stable flow and is then quickly slam shut with the wing valve for a period. Pressure will build-up (PBU) in the well and PBU test will be available (Fig. 5.15).

The production rate and bottom-hole pressure time relevant to the buildup test are shown in Figs. 5.16 and 5.17. Where t_p denotes the production time and t represents the time since the shut-in. Before shutting in the well, the pressure is measured; thereafter, the wellbore pressure is quantitatively recognized to determine the reservoir parameter values and the wellbore condition.

Before starting the well test interpretation, much important information must be available and measured accurately such as, choke size variation, tubing size, casing sizes, and tested interval depth. This information has a significant impact on the interpretation process (Tarek 2001). Well stabilization at a constant rate is also

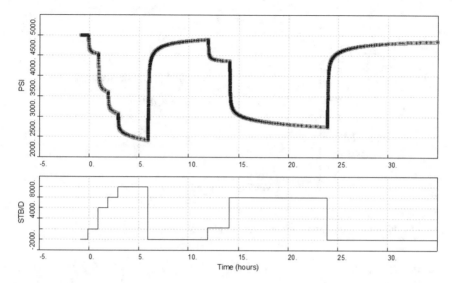

Fig. 5.14 Show well test period for an appraisal well

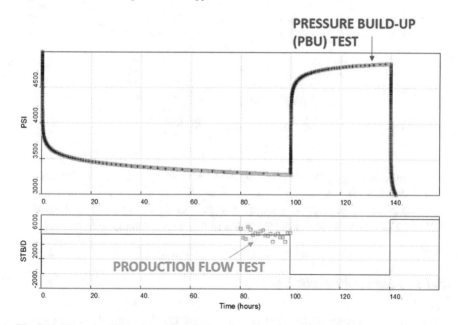

Fig. 5.15 Show the BU test is performed when the well has a stable flow rate period

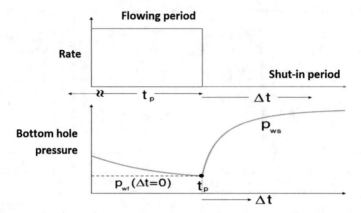

Fig. 5.16   Flowrate and pressure behaviour for an ideal BU test

Fig. 5.17   Horner Plot for a
BU test

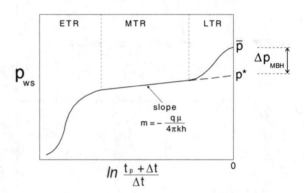

crucial, otherwise applying conventional methods to interpret the test results can lead
to widely erroneous results.

Many methods are offered for the interpretation of the BU test, such as the Horner
plot and MDH plot method, but the Horner plot is the commonly used method.

5.4.1   Horner Plot Method

This approach assumes that the reservoir is infinite in extent and a small amount of
hydrocarbons have been produced from the well during the drawdown period before
the shut-in of the well (Dake 1977). For the duration of an infinite-acting time, the
pressure profile can be indicated as follows:

$$p_{ws(\Delta t)} = p_i - \frac{q_s B \mu}{2\pi k h} \times \frac{1}{2}\left(\ln\left(t_p + \Delta t_D\right) - \ln \Delta t_D\right) \qquad (5.16)$$

Can be simplified as:

$$p_{ws(\Delta t)} = p_i - \frac{q_s B \mu}{2\pi kh} \times \ln\left(\frac{t_p + \Delta t}{\Delta t}\right) \tag{5.17}$$

In filed unit:

$$p_{ws(\Delta t)} = p_i - \frac{162.6 q_s B \mu}{kh} \times \log\left(\frac{t_p + \Delta t}{\Delta t}\right) \tag{5.18}$$

The linear relationship in Horner's plot is characterized between p_{ws} and $\ln\left(\frac{t_p + \Delta t}{\Delta t}\right)$. This equation indicates also that shut-in bottom-hole pressure can reach the initial pressure p_i.

$$p_{ws} = m \times \ln\left(\frac{t_p + \Delta t}{\Delta t}\right) + p^* \tag{5.19}$$

where the slope m can be expressed as:

$$m = -\frac{q_s B \mu}{4\pi kh} \tag{5.20}$$

In field unit:

$$m = -162.6 \frac{q_s B \mu}{kh} \tag{5.21}$$

The intercept, p^*, can be referred to initial reservoir pressure.

$$p^* = p_i \tag{5.22}$$

Using Eq. 5.20, permeability can be expressed as:

$$k = -\frac{q_s B \mu}{4\pi mh} \tag{5.23}$$

By using a semi-log plot, the flow capacity, (kh), can be obtained via the slope of the build-up (Ronald 1990).

$$kh = -\frac{q_s B \mu}{4\pi m} \tag{5.24}$$

As mentioned in the upper section, to estimate the skin factor the pressure values before shut-in the well is measured. The flowing pressure in infinite-acting reservoirs before the shut-in well is defined as follows:

$$p_{wf\,(\Delta t=0)} = p_i - \frac{q_s B \mu}{4\pi k h} \times \left(\ln \frac{k t_p}{\varnothing \mu C_t r_w^2} + 0.80908 + 2S \right) \qquad (5.25)$$

By replacing p_i with p^* and simplifying the slope m, the equation can be written as:

$$p_{wf(\Delta t=0)} = p^* - m \times \left(\ln \frac{k t_p}{\varnothing \mu C_t r_w^2} + 0.80908 + 2S \right) \qquad (5.26)$$

The skin factor can be estimated using the following equation:

$$S = \frac{1}{2} \left(\frac{p_{wf(\Delta t=0)} - p^*}{m} - \ln \frac{k t_p}{\varnothing \mu C_t r_w^2} - 0.80908 \right) \qquad (5.27)$$

In field unit:

$$S = 1.153 \left[\frac{p_{wf(\Delta t=0)} - p^*}{m} - \log_{10} \frac{k t_p}{\varnothing \mu C_t r_w^2} + 3.2275 \right] \qquad (5.28)$$

Once limited data points are obtained, extrapolating the line to get p^* might not be correct. But it's possible to determine the pressure just one hour after closing the well as shown in Fig. 5.18.

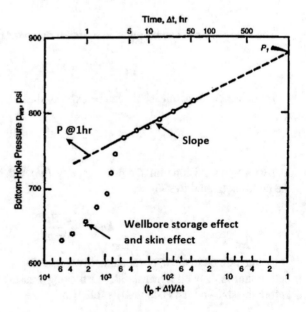

Fig. 5.18 Horner plot

In this case, the shut-in pressure can be written as:

$$p_{ws} = m \log\left(\frac{t_p + \Delta t}{\Delta t}\right) + p^* \qquad (5.29)$$

$$p_{wf\,(\Delta t=0)} = p_{1h} = p^* + m \, \log(t_p + 1) \qquad (5.30)$$

The final equation can be written as:

$$p^* = p_{1h} - m \log(t_p + 1) \qquad (5.31)$$

The final skin factor equation can be written as:

$$S = 1.153 \left[\frac{p_{wf\,(\Delta t=0)} - p_{1h}}{m} + \log\left(\frac{t_p + 1}{t_p}\right) - \log\frac{k}{\emptyset \mu C_t r_w^2} + 3.2275\right] \qquad (5.32)$$

5.4.1.1 Example 5.5

The buildup test results for Well-S-33 are displayed in Table 5.5. From a well placed in the centre of a square-shaped reservoir, the following reservoir parameters were obtained (Tiab 1993):

$$r_w = 4\,\text{in}, \quad h = 44\,\text{ft}, \quad \phi = 12\%$$
$$\mu = 0.76\,\text{cp}, \quad B = 1.24\,\text{RB/STB}, \quad N_p = 4550\,\text{STB}$$
$$A = 40\,\text{acres}, \quad q = 340\,\text{BPD}, \quad c_t = 36 \times 10^{-6}\,\text{psia}^{-1}$$
$$P_{wf} = 2980\,\text{psia}$$

It is required to determine reservoir permeability and skin factor, and the average reservoir pressure.

Solution
Find t_p with;

$$t_p = \frac{24N_p}{q} = \frac{24 \times 4550}{340} = 321.176\,\text{h}$$

Table 5.5 Buildup data	Δt, h	P_{ws}, psia
	0	2980
	0.1	3100
	0.2	3150
	0.3	3200
	0.5	3250
	0.75	3275
	1	3290
	2	3315
	3	3325
	4	3330
	5	3335
	7	3342
	10	3350
	15	3360
	20	3364
	30	3370
	40	3372
	50	3374
	60	3375
	70	3376
	80	3377

Calculate the Horner time, $(t_p + \Delta t)/\Delta t$, to each pressure value. The measured pressure data is presented in Table 5.6 and builds the Horner plot given in Fig. 5.19. From the Horner plot, the slope and intercept are read to be 44 psia/cycle and 3306 psia respectively.

Find permeability;

$$K = \frac{162.6 q \mu B}{mh} = \frac{162.6 \times 340 \times 0.76 \times 1.24}{44 \times 44} = 26.9 \, \text{mD}$$

Find skin factor;

$$S = 1.153 \left[\frac{p_{wf(\Delta t=0)} - p_{1h}}{m} + \log\left(\frac{t_p + 1}{t_p}\right) - \log\frac{k}{\emptyset \mu C_t r_w^2} + 3.2275 \right]$$

$$S = 1.1513 \left[\frac{3306 - 2980}{44} - \log\left(\frac{26.91}{0.12 \times .76 \times 36 \times 10^{-6} \times (0.333)^2}\right) + 3.2275 \right]$$

$$S = 3.18$$

Table 5.6 Horner time for each pressure value

Δt, h	P_{ws}, psia	$(t_p + \Delta t)/\Delta t$	$(t_{pss} + \Delta t)/\Delta t$	ΔP, psia	$t * \Delta P'$, psia
0	2980			0	0
0.1	3100	3213.00	807.450	120	83.41
0.2	3150	1607.00	404.225	170	100.23
0.3	3200	1071.67	269.817	220	110.92
0.5	3250	643.40	162.290	270	85.89
0.75	3275	429.27	108.527	295	59.48
1	3290	322.20	81.645	310	41.84
2	3315	161.60	41.323	335	34.48
3	3325	108.07	27.882	345	22.35
4	3330	81.30	21.161	350	20.29
5	3335	65.24	17.129	355	21.45
7	3342	46.89	12.521	362	21.96
10	3350	33.12	9.065	370	23.97
15	3360	22.41	6.376	380	21.42
20	3364	17.06	5.032	384	14.87
30	3370	11.71	3.688	390	12.65
40	3372	9.03	3.016	392	9.02
50	3374	7.42	2.613	394	8.47
60	3375	6.35	2.344	395	7.14
70	3376	5.59	2.152	396	8.55
80	3377	5.02	2.008	397	9.76

5.4.2 MDH Plot Method

MDH method was presented by Miller-Dye-Hutchinson in 1950 (Xiao and Reynolds 1992). This method determines the average reservoir pressure in closed circular or square drainage areas by using the plotted test data (P_{ws} versus log Δt). This method is used only for wells flowing at semi or pseudo-steady state conditions before conducting the buildup test. To use MDH, select any fitting time on the semi-log straight line plot, Δt, and read the equivalent pressure, P_{ws}. After that the dimensionless shut-in time based on the drainage area can be determined along with the reservoir pressure using the below equations:

$$\Delta t_{DA} = \left(\frac{0.0002637 K}{\emptyset \mu C_t A} \right) \Delta t \tag{5.33}$$

$$\overline{P} = P_{ws} + \frac{m P_{DMDH(\Delta t_{DA}) upper\ curve}}{1.1513} \tag{5.34}$$

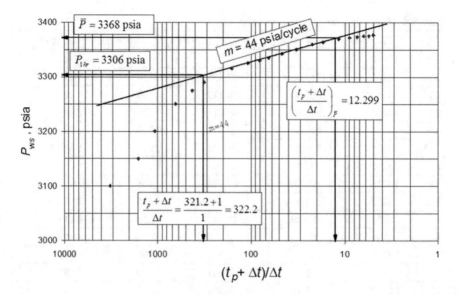

Fig. 5.19 Buildup test data for well-S-33, horner plot

$$P_i = P_{ws} + \frac{m\,P_{DMDH(\Delta t_{DA})Lower\ curve}}{1.1513} \tag{5.35}$$

where Δt and its corresponding P_{ws} are read from the straight-line portion of the MDH plot and PDMDH is obtained from MDH chart in Fig. 5.20.

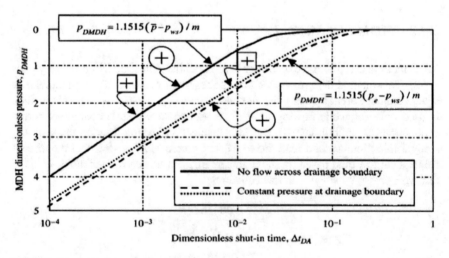

Fig. 5.20 Dimensionless pressure for square reservoir

5.4.2.1 Example 5.6

Using the MDH plot, Fig. 5.21, choose any convenient point on the semi-log straight line and calculate the dimensionless shutting-in time using the chosen time.

Solution
From Fig. 5.19, $\Delta t = 10$ h and $(P_{ws}) = 3350$ psia were chosen.

$$\Delta t_{DA} \left(\frac{0.0002637 K}{\emptyset \mu C_t A} \right) \Delta t = \frac{0.0002637 \times 26.91}{0.12 \times 0.76 \times 36 \times 10^{-6} \times 40 \times 43560} \times 10$$

$$= 0.0124$$

Find the average reservoir pressure, by using the obtained Δt_{DA} and assume P_{DMDH} equal to 0.6 for a no-flow boundary square reservoir.

$$\overline{P} = P_{ws} + \frac{m \, P_{DMDH(\Delta t_{DA}) Lower \, curve}}{1.1513} = 3350 + \frac{44 \times 0.6}{1.1513} = 3372.9 \, \text{Psia}$$

5.5 Multiple Well Testing

These are used to connect wells and determine the properties of the inter-well reservoir. The multiple-well testing principle can also be applied to different sets of

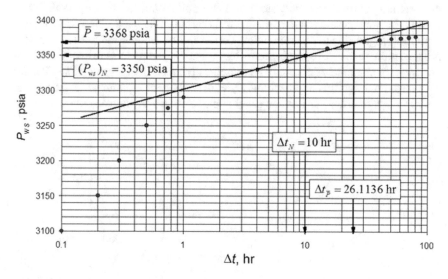

Fig. 5.21 MDH plot

perforations in the same wellbore. Multiple-well tests between offsetting wells determine reservoir properties on an area-by-area basis. Vertical reservoir permeability is usually determined by tests performed between different sets of perforations in the same wellbore. Multiple-well tests detect reservoir heterogeneity better than single-well tests. Multiple-well tests are classified into two types: interference and pulse.

5.5.1 Interference Test

As a static observation well's pressure response is being monitored, a production or injection well's flow rate is abruptly altered (Fig. 5.22). The degree of communication between the two wells may be determined by measuring the time it takes for the pressure transient to reach the observation well and the resulting pressure change.

   where

$$r_{inf} = 0.029 \sqrt{\frac{kt}{\emptyset \mu C_t}} \tag{5.36}$$

Well communication between offset wells is a typical issue, especially in high-permeability reservoirs. An interference test is a technique for figuring out how wells are connected to one another and evaluating the interaction's impact on surrounding well output. As mentioned above, in this type of test, one well has flowed while another adjacent well's pressure response is recorded. If the wells can communicate, the adjoining shut-in well will see a pressure pulse from the flowing well. When considering field development spacing, this is helpful. When a well is closed off, it is used in storage fields to determine field deliverability. An interference test can also produce more realistic reservoir parameters when the parameters are heterogeneous. Porosity, for example, is normally measured using open-hole logs or (if applicable) core data, although this approach only yields porosity at a particular region in the reservoir. An interference test can provide the relative porosity between two wells,

Fig. 5.22 Influence region for interference

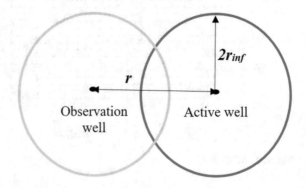

which might also differ from what logs or cores reported. Because porosity is a crucial determinant in reservoir volume, determining average porosity more accurately will result in more accurate volumetric estimations.

5.5.2 Pulse Tests

A substitute for interference tests where there is little physical space between wells, as in pattern floods. By alternately flowing and closing in the active well, a sequence of pressure pulses is produced rather than just adjusting the rate. One or more observation wells are used to measure the pressure. In reservoirs with broad well spacing that have low permeability and high compressibility, this type of test is challenging to successfully conduct.

Interference test is normally much more expensive than pulse test due to the revenue loss caused by having to shut down a significant part or all of the tested reservoir to perform the test. There is also ambiguity in interference test interpretation because it is unknown whether an observed response was caused by the active well.

In a pulse test (Fig. 5.23), if a repeated signal is received in an observation well, there is little uncertainty that it was produced by the rate changes in the active well. During a pulse test, a coded signal is introduced into the reservoir via rate variations at the source well, and the pressure response is recorded at the observation well.

The pulse test provides directional data regarding transmissibility and storativity.

$$\text{Transmissibility (T)} = \frac{kh}{\mu} \tag{5.37}$$

$$\text{Storativity (S)} = \emptyset C_t h \tag{5.38}$$

Lag time and pulse height are dependent on T/S and T, respectively. S may provide details regarding the effective reservoir thickness transmitting pulse. The T/S ratio is the single output of an interferenc test.

5.5.2.1 Benefits of Pulse Test

- Evaluation of reservoir properties between wells with more certainty than interference tests.
- Interpretation is done with greater confidence than interference test when background or unknown interference is present.

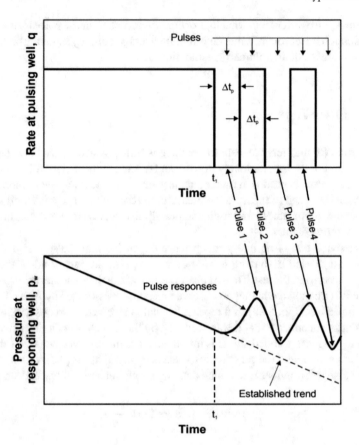

Fig. 5.23 Pulse test

5.6   Closed-Chamber Test

A closed-chamber test is carried out in the borehole using the drill string (Fig. 5.24). The surface valve is close for the period of the test. It is suggested to use a downhole gauge. The air in the string is compressed as the well begins to flow, and the volume of reservoir-fluid inflow as a function of time is estimated by monitoring the surface pressure. The downhole valve is shut-in to stop flow when the surface pressure reaches a predetermined level. This guarantees the production of a specific volume of reservoir fluid. No hydrocarbons are brought to the surface. Injection of fluids into the drill or completion string. Closed chamber tests are environmentally friendly and safe when H_2S is present (Xiao and Reynolds 1992).

The drill stem test sequence for the closed chamber test is scheduled to last 4 to 5 h. The scope and goals of the test are identical to those for any short-term drill stem test using mechanical gauges.

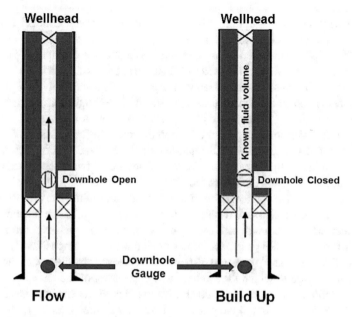

Fig. 5.24 Well schematic for closed-chamber test

5.6.1 Application of Closed Chamber Testing

Back surge completion testing is the application of closed chamber testing. The following is a summary of a transient pressure analysis method for the closed chamber well test. Using log–log type curve matching, the method enables the determination of reservoir transmissibility and well bore skin. The technique is thought to be useful in the analysis of back surge data for perforation cleaning. The radial diffusivity equation's solution is found by superimposing the cumulative influx, constant pressure solution. Superposition makes it easier to take into account complex well bore geometry and non-ideal chamber gas behaviour while avoiding many of the drawbacks of direct solution.

There is a sensitivity analysis of the tool and reservoir parameters. The findings show that, in contrast to the slug test, tool geometry significantly affects the closed chamber test's dimensionless pressure response.

To extend the time that the closed chamber test functions as a slug test, tool design processes are established. These processes are suitable for analysis by type curve matching with published slug test data. The superposition results further demonstrate that the bottom hole pressure response is not dramatically altered by assuming ideal chamber gas behaviour. To mimic the pressure response of a closed chamber well test, a computer model was built. The radial diffusivity equation's superposition of the constant pressure, and cumulative inflow solution was employed in the model. The model also took into account the impacts of real gas compressibility, but it ignored the effects of friction and momentum. When creating the mathematical model, isothermal chamber gas compression was assumed.

Ramey (1985), conducted a slug test which is a specific example of the general closed chamber test, which was used to assess the superposition model's capacity to replicate their findings. To create standards for the design and analysis of the closed chamber test, a tool and reservoir parameter sensitivity research was carried out. The outcomes of simulating the closed chamber test by superimposing the cumulative inflow solution with constant pressure support the following conclusions: The Ramey 1985 slug test results, which also disregarded momentum and friction factors, can be replicated using the superposition model. The difference between the closed chamber test and the slug test was demonstrated. The ratio of the initial to the final well bore storage determines the change in logarithmic coordinates of the late time dimensionless closed chamber pressure response. The bottom hole pressure response of the closed chamber test is unaffected by non-ideal chamber gas behaviour at moderate reservoir pressure (5000 psig). The effect of chamber gas composition is therefore minimal. The bottom hole pressure response of the closed chamber test is unaffected by the temperature at which the isothermal compression of the chamber gas takes place throughout a range of 100–500 (F). If the effect of the chamber gas compression is more than or equal to the effect of the slug test, then a larger portion of the closed chamber test response will be equivalent to a slug test and hence acceptable for slug test type curve analysis, if during the test the impact of the chamber gas compression is kept to a minimum. According to the sensitivity analysis, the effect of chamber gas compression will be diminished by increasing the chamber's length, diameter, and beginning fluid column length. To prevent deviance from the corresponding early time slug test response, an initial chamber gas pressure close to the atmosphere is needed. It is clear from the late-time dimensionless closed chamber type curves produced by the superposition model that pressure measurement within the first 60 s of the flow period will be necessary to assess skin impact for wells with more or fewer flow capacities reaching 1000 md-ft.

5.7 Summary

The majority of well tests include adjusting the rate and analyzing the pressure change caused by the rate variation. To perform a well test successfully, one must be able to monitor time, rate, pressure, and adjust the rate. Well tests can be used to determine reservoir properties if they are appropriately designed. Most tests allow a certain amount of fluid to flow from or into a reservoir. The well is subsequently closed, and pressures are measured as the formation equilibrates. In this chapter many types of well testing were covered in detail, including buildup and drawdown tests (horner plot and MDH plot) as a function of constant rate production and smoothly changing rate.

References

Ahmed Tarek, "Reservoir Engineering Handbook", 2001.

Dake L.P., "Fundamentals of Reservoir Engineering", 1977.

Ramey, H.J., Jr. 1985. "Advances in Well Test Analysis," Invited lecture National Conference of Petroleum Independents, Pittsburgh, PA, May 21–23, 1985. Proceedings.

Ronald N.: "Modern Well Testing Analysis", 1990.

Streltsova, T.D. and McKinley, R.M., "Effect of Flow Time Duration on BuildUp Pattern for Reservoirs with Heterogeneous Properties", 1984.

Tiab D. PE-5553: Well Test analysis. Lecture notes. The University of Oklahoma; 1993

Xiao, Jinjiang, and A.C. Reynolds. "New Methods for the Analysis of Closed-Chamber Tests." Paper presented at the SPE Western Regional Meeting, Bakersfield, California, March 1992. https://doi.org/10.2118/24059-MS.

Chapter 6
The Principle of Superposition

6.1 Overview

The well may be flow for a while during well tests, and then "shut-in". To gather the information that can be utilized to ascertain different reservoir qualities, it may also be flowing at a series of varying rates. A complicated pressure signal is produced by these sequences. This chapter will introduce the concept of superposition and demonstrate how it may be applied to the development of techniques for the analysis of these intricate pressure signals, enabling us to determine the values of critical reservoir characteristics like permeability and storativity (Roland 1995; John 1982; Sabet 1991).

6.2 Principle of Superposition and Constraints

The radial diffusivity equation's solutions it seems to be appropriate only for representing the pressure distribution in an infinite reservoir produced by continual production from a single well. To examine the fluid flow during the unsteady-state flow period, the best general approach is required because real reservoir systems typically have multiple wells working at various rates. The principle of superposition is a significant idea that can be used to avoid any constraints placed on different approaches to solving the transient flow equation. According to the superposition principle, the diffusivity equation may contain any sum of its component solutions. Applying this idea will allow you to take into consideration the transient flow solution's impacts from multiple wells, rate changes, boundary changes, and pressure changes, among others. Slider (1976) provided an outstanding overview and explanation of the real-world uses of the superposition principle.

© The Author(s), under exclusive license to Springer Nature Switzerland AG 2023   137
T. A-A. O. Ganat, *Modern Pressure Transient Analysis of Petroleum Reservoirs*,
Petroleum Engineering, https://doi.org/10.1007/978-3-031-28889-0_6

6.3 Multiple-Well Cases

If we first begin producing oil from a reservoir at a uniform pressure pi, then will distort the pressure profile at the wellbore, whose slope is determined by Darcy's Law (Fig. 6.1). The diffusivity equation will explain how soon this distortion will develop inside the reservoir. The bending of the pressure profile in the case of a drawdown is described as concave. The pressure will decrease over the whole production phase, with the profile being most concave close to the well. As more and more fluid is generated from deeper inside the reservoir, the concavity surrounding the well will gradually become less noticeable.

The easiest way to understand superposition is to calculate the pressure drop at a particular location in a field where two sinks are placed. Take the three-well infinite system in Fig. 6.2 as an example. Well-1 begins to produce at rate q_1, and well-2 begins to produce at rate q_2, at time $t = 0$. The pressure at well-3, the shut-in observation pressure, should be recorded. This is accomplished by combining the pressure change at well-3 induced by well-1 with the pressure change at well-3 caused by well-2 (Fig. 6.3).

$$P_{Total\ Drop\ at\ well\#1} = P_{Drop\ due\ to\ well\#1} + P_{Drop\ a\ due\ to\ well\#2} + P_{Drop\ due\ to\ well\#3} \tag{6.1}$$

For instance, the pressure drops at well-1 (Fig. 6.4) due to its production is given by the log approximation to the Ei-function solution presented by Eq. (6.2):

$$\left(p_i - p_{wf}\right) = (\Delta p)_{well\#1} = \frac{162.6\, q_1 B\mu}{kh}\left[\log\left(\frac{kt}{\emptyset\mu C_t r_w^2}\right) - 3.23 + 0.869\, S\right] \tag{6.2}$$

Since the log approximation cannot be used to calculate the pressure at a great distance r from the well when $x > 0$, the extra pressure drops at well-1 caused by the production from wells 2 and 3 must be expressed in terms of the Ei-function solution. Therefore:

Fig. 6.1 Drawdown sink

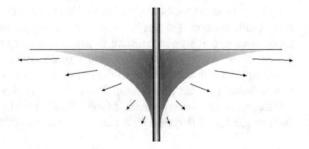

Fig. 6.2 Principle of superposition

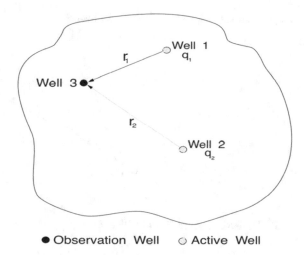

Fig. 6.3 Illustrate pressure change at the observation well

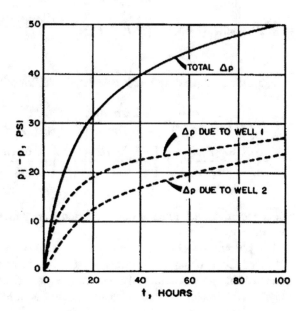

$$(p_i - p_{wf}) = (\Delta p)_{well\#1} = \frac{162.6q_1 B\mu}{kh}\left[\log\left(\frac{kt}{\emptyset\mu C_t r_w^2}\right) - 3.23 + 0.869\,S\right] \tag{6.3}$$

$$p(r,t) = p_i + \left[\frac{70.6\,q\,B\mu}{kh}\right] \times E_i\left[\frac{-948\emptyset\mu C_t r_w^2}{kt}\right] \tag{6.4}$$

To determine the extra pressure drop caused by the two wells, use the following formula:

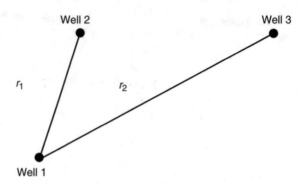

Fig. 6.4 Wells layout

$$(\Delta p)_{Drop\ due\ to\ well\#2} = p_i - p(r_1, t) = -\left[\frac{70.6\,q_1 B \mu}{kh}\right] \times E_i\left[\frac{-948\emptyset\mu C_t r_1^2}{kt}\right]$$

(6.5)

$$(\Delta p)_{Drop\ due\ to\ well\#3} = p_i - p(r_2, t) = -\left[\frac{70.6\,q_2 B \mu}{kh}\right] \times E_i\left[\frac{-948\emptyset\mu C_t r_2^2}{kt}\right]$$

(6.6)

Next, the total pressure loss is calculated as follows:

$$\left(p_i - p_{wf}\right)_{total\ at\ well\#1} = \frac{162.6\,q_1 B \mu}{kh}\left[\log\left(\frac{kt}{\emptyset\mu C_t r_w^2}\right) - 3.23 + 0.869\,S\right]$$
$$-\left[\frac{70.6\,q_2 B \mu}{kh}\right] \times E_i\left[\frac{-948\emptyset\mu C_t r_1^2}{kt}\right] - \left[\frac{70.6\,q_3 B \mu}{kh}\right] \times E_i\left[\frac{-948\emptyset\mu C_t r_2^2}{kt}\right]$$

(6.7)

where the production rates of wells 1, 2, and 3 are indicated as Qo_1, Qo_2, and Qo_3, respectively.

The pressure at wells 2 and 3 may be determined using the computational method described above. It may also be expanded to encompass any quantity of wells running in an unstable state. Additionally, it should be highlighted that the skin factors must only be specified for the well in question if the site of interest is an active well.

6.3.1 Example 6.1

Assume that during 15 h, the three wells in Fig. 6.5 are producing under a transient flow scenario. The following are the well data given:

$Qo1 = 100\,STB/day$, $Qo2 = 160\,STB/day$, $Qo3 = 200\,STB/day$,

$pi = 4500\,psi$, $Bo = 1.\,20\,bbl/STB$, $ct = 20 \times 10^{-6}\,psi^{-1}$, $(s)well1 = -0.\,5$,

Fig. 6.5 Three wells layout

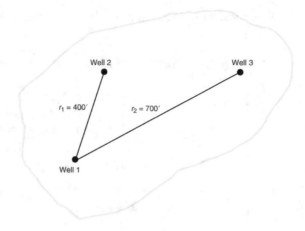

h = 20 ft, φ = 15%, k = 40 md, rw = 0.25 ft, μo = 2.0 cp,
r1 = 400 ft, r2 = 700 ft.

If the three wells are producing at a constant flow rate, calculate the sand face flowing pressure at well 1.

Solution

1. Determine the pressure drop at well-1 due to its production:

$$\left(p_i - p_{wf}\right) = (\Delta p)_{well\#1} = \frac{162.6\,q_1\,B\mu}{kh}\left[\log\left(\frac{kt}{\emptyset\mu C_t r_w^2}\right) - 3.23 + 0.869\,S\right]$$

$$(\Delta p)_{well\#1} = \frac{162.6 \times 100 \times 1.2 \times 2}{40 \times 20}$$

$$\left[\log\left(\frac{40 \times 15}{0.15 \times 2 \times 20 \times 10^{-6} \times (0.25)^2}\right) - 3.23 + 0.869 \times 0\right] = 270\,\text{Psia}$$

2. Determine the pressure drop at well-1 as a result of well-2 production:

$$(\Delta p)_{Drop\ due\ to\ well\#2} = P_i - p(r_1, t) = -\left[\frac{70.6q_1 B\mu}{kh}\right] \times E_i\left[\frac{-9480\mu C_t r_1^2}{kt}\right]$$

$$(\Delta p)_{Drop\ due\ to\ well\#2} = -\left[\frac{70.6 \times 160 \times 1.2 \times 2}{40 \times 20}\right]$$

$$\times E_i\left[\frac{-948 \times 0.15 \times 2 \times 20 \times 10^{-6} \times (400)^2}{40 \times 15}\right]$$

$$= 33.89\left[-E_i(-1.517)\right]$$

$$= 33.89 \times 0.13 = 4.4\,\text{Psia}$$

3. Determine the pressure drop at well-1 as a result of well-2 production:

$$(\Delta p)_{Drop\ due\ to\ well\#3} = p_i - p(r_2, t) = -\left[\frac{70.6\,q_2 B\mu}{kh}\right] \times E_i\left[\frac{-948\emptyset\mu C_t r_2^2}{kt}\right]$$

$$(\Delta p)_{Drop\ due\ to\ well\#3} = -\left[\frac{70.6 \times 200 \times 1.2 \times 2}{40 \times 15}\right]$$

$$\times\ E_i\left[\frac{-948\ \times 0.15 \times 2 \times 20 \times 10^{-6} \times (700)^2}{40 \times 15}\right]$$

$$= -\ 42.4\ \times\ E_i\ [-4.65]$$

$$= -\ 42.4 \times\ 1.84 \times\ 10^{-3}$$

$$= 0.08\ \text{Psia}$$

4. Determine the total pressure drop caused at well-1:

$$\left(p_i - p_{wf}\right)_{total\ at\ well\#1} = \frac{162.6\,q_1 B\mu}{kh}\left[\log\left(\frac{kt}{\emptyset\mu C_t r_w^2}\right) - 3.23 + 0.869\,S\right]$$

$$-\left[\frac{70.6\,q_2 B\mu}{kh}\right] \times E_i\left[\frac{-948\emptyset\mu C_t r_1^2}{kt}\right] - \left[\frac{70.6 q_3 B\mu}{kh}\right] \times E_i\left[\frac{-948\emptyset\mu C_t r_2^2}{kt}\right]$$

$$\left(p_i - p_{wf}\right)_{total\ at\ well\#1} = 270 + 4.4 + 0.08 = 274.5\,\text{Psia}$$

5. Determine Bottom hole pressure at well-1:

$$P_{wf} = 4500 - 274.5 = 4225.5\,\text{Psia}$$

6.4   Effects of Fluctuating Flow Rates

Throughout the transient flow periods, the wells must continue to produce at a constant rate. It must be able to forecast how the pressure will behave when the rate varies because practically every well produces at a fixed rate. According to Superposition, "any change in flow rate in a well will result in a pressure response that is independent of the pressure responses caused by the previous rate changes." The overall pressure drop that has occurred at any given time is thus equal to the sum of the individual pressure changes induced by each net flow rate change at that time.

Take into account the scenario of a shut-in well, or $Q = 0$, which was then permitted to produce at a succession of consistent rates during the various periods depicted in Fig. 6.6. To determine the total pressure drop at the sand face at time t4, the distinct constant-rate solutions are combined at the given rate-time sequence to generate the composite solution, or:

Fig. 6.6 Well production
and pressure history

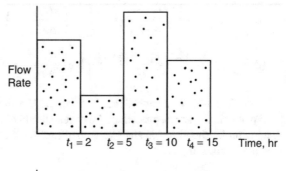

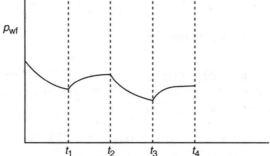

$$(\Delta P)_{total} = (\Delta P)_{due\ to\ (Q_o1-0)} + (\Delta P)_{due\ to\ (Q_o2-Q_o1)}$$
$$+ (\Delta P)_{due\ to\ (Q_o3-Q_o2)} + (\Delta P)_{due\ to\ (Q_o4-Q_o3)} \qquad (6.8)$$

According to the aforementioned statement, the four different flow rates each make up four contributions to the overall pressure drop:

The first contribution is the outcome of raising the rate from 0 to Q_1 and it applies for the full time t_4, so:

$$(\Delta P)_{q1-0} = \frac{162.6(q_1 - 0)B\mu}{kh}\left[\log\left(\frac{kt}{\emptyset \mu C_t r_w^2}\right) - 3.23 + 0.869\,S\right] \qquad (6.9)$$

It is crucial to pay attention to how the rate in the equation above has changed, or (new rate–old rate). The pressure disturbance is brought on by the rate change. It should also be emphasized that the term "time" in the equation refers to the entire period that has passed since the rate adjustment took effect.

The second influence results from decreasing the rate from Q_1 to Q_2 at t_1, therefore:

$$(\Delta P)_{q2-q1} = \frac{162.6(q_2 - q_1)B\mu}{kh}\left[\log\left(\frac{k(t_4 - t_1)}{\emptyset \mu C_t r_w^2}\right) - 3.23 + 0.869\,S\right] \qquad (6.10)$$

Using the same concept, the following calculation can be made for the two extra contributions from Q_2 to Q_3 and Q_3 to Q_4:

$$(\Delta P)_{q_3-q_2} = \frac{162.6(q_3 - q_2)B\mu}{kh}\left[\log\left(\frac{k(t_4 - t_2)}{\emptyset\mu C_t r_w^2}\right) - 3.23 + 0.869\ S\right] \quad (6.11)$$

$$(\Delta P)_{q_4-q_3} = \frac{162.6(q_4 - q_3)B\mu}{kh}\left[\log\left(\frac{k(t_4 - t_3)}{\emptyset\mu C_t r_w^2}\right) - 3.23 + 0.869\ S\right] \quad (6.12)$$

To depict a well with several rate fluctuations, the aforementioned approach might be expanded. The stated procedure, however, should only be used if the well has been flowing under the unstable state flow condition for the whole time since it began to flow at its initial rate.

6.4.1 Example 6.2

Figure 6.7 displays the rate history of a well-R36 that has been flowing for 15 h with transient flow conditions. Applying the provided well data, calculate the well bottom hole pressure after 15 h.

$$p_i = 5000\,\text{psi},\ h = 20\,\text{ft},\ B_o = 1.1\,\text{bbl/STB}$$
$$\varphi = 15\%,\ \mu_o = 2.5\,\text{cp},\ r_w = 0.3\,\text{ft}$$
$$c_t = 20 \times 10^{-6}\,\text{psi}^{-1},\ S = 0,\ k = 40\,\text{md}$$

Fig. 6.7 Production rate history of a well-R36

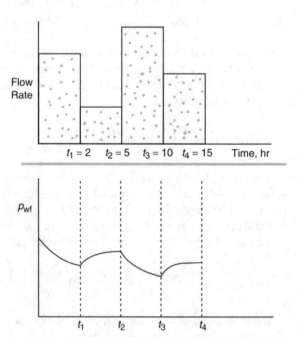

Solution

1. Determine the pressure drop resulting from the initial flow rate throughout the full flow period:

$$(\Delta P)_{q1-0} = \frac{162.6(q_1 - 0)B\mu}{kh}\left[\log\left(\frac{kt}{\emptyset\mu C_t r_w^2}\right) - 3.23 + 0.869\,S\right]$$

$$(\Delta P)_{q1-0} = \frac{162.6(100 - 0)1 \times 2.5}{40 \times 20}$$

$$\left[\log\left(\frac{40 \times 15}{0.15 \times 2.5 \times 20 \times 10^{-6} \times (0.3)^2}\right) - 3.23 + 0.869 \times 0\right] = 320\,\text{Psia}$$

2. Determine the extra pressure change caused by a decrease in flow rate from 100 to 70 STB/day:

$$(\Delta P)_{q2-q1} = \frac{162.6(q_2 - q_1)B\mu}{kh}\left[\log\left(\frac{k(t4 - t_1)}{\emptyset\mu C_t r_w^2}\right) - 3.23 + 0.869\,S\right]$$

$$(\Delta P)_{q2-q1} = \frac{162.6(70 - 100) \times 1.1 \times 2.5}{40 \times 20}$$

$$\left[\log\left(\frac{40(15 - 2)}{0.15 \times 2.5 \times 20 \times 10^{-6} \times (0.3)^2}\right) - 3.23 + 0.869 \times 0\right]$$

$$= 95\ \text{Psia}$$

3. Determine the extra pressure change caused by the change of the flow rate from 70 to 150 STB/day:

$$(\Delta P)_{q3-q2} = \frac{162.6(q_3 - q2)B\mu}{kh}\left[\log\left(\frac{k(t4 - t2)}{\emptyset\mu C_t r_w^2}\right) - 3.23 + 0.869\,S\right]$$

$$(\Delta P)_{q3-q2} = \frac{162.6(150 - 70)1 \times 2.5}{40 \times 20}$$

$$\left[\log\left(\frac{40(15 - 5)}{0.15 \times 2.5 \times 20 \times 10^{-6} \times (0.3)^2}\right) - 3.23 + 0.869 \times 0\right] = 249\,\text{Psia}$$

4. Determine the extra pressure change caused by the change of the flow rate from 150 to 85 STB/day:

$$(\Delta P)_{q4-q3} = \frac{162.6(q_4 - q3)B\mu}{kh}\left[\log\left(\frac{k(t4 - t3)}{\emptyset\mu C_t r_w^2}\right) - 3.23 + 0.869\,S\right]$$

$$(\Delta P)_{q4-q3} = \frac{162.6(85 - 150)1.1 \times 2.5}{40 \times 20}$$

$$\left[\log\left(\frac{40 \times (15 - 10)}{0.15 \times 2.5 \times 20 \times 10^{-6} \times (0.3)^2}\right) - 3.23 + 0.869 \times 0\right]$$

$$= 190 \text{ Psia}$$

5. Determine the total pressure drop:

$$(\Delta P)_{total} = 320 + (-95) + 249 + (-190) = 284 \text{ Psia}$$

6. Determine the bottom hole pressure after 15 h of transient flow:

$$P_{wf} = 5000 - 284 = 4716 \text{ Psia}$$

6.5   Reservoir Boundary Effects

Predicting the pressure of a well in a bounded reservoir is another use of the superposition principle. As an example, Fig. 6.8, displays a well that is placed L meters away from the sealing fault or other non-flow barrier. The pressure gradient expression shown here can be used to illustrate the no-flow boundary:

$$\left(\frac{\partial p}{\partial l} \right)_{boundry} = 0 \tag{6.13}$$

Understanding the impact of boundaries might be just as crucial as quantitatively analyzing the test. It is challenging to identify this problem since different reservoir models could provide equivalent pressure responses. It was imperative that the model used to quantitatively assess the test be compatible with interpretations based on geological and geophysical data. Type-curve matching or regression analysis using recent well-test analysis software may be a relatively straightforward test analysis technique after the suitable reservoir model has been determined.

For the majority of tests, the early- and middle-time portions of the diagnostic plots for a buildup test and a drawdown test are nearly identical. However, late in buildup and drawdown tests, boundary effects might result in rather varied forms for the same reservoir model. The usual practice of analyzing buildup tests on drawdown type curves using "equivalent time" functions adds to this problem, (Radial flow, linear flow, and bilinear flow all have various corresponding time functions.).

The superposition method, which considers boundary effects, is commonly known as the approach of images. The issue with the system configuration depicted in Fig. 1.30 is thus reduced to understanding how the image well impacts the actual well. Either the total pressure drops at the real well will equal the sum of the pressure drop caused by the well's production plus the excessive pressure drop caused by a comparable well two L away, or:

$$(\Delta P)_{total} = (\Delta P)_{actual \, well} + (\Delta P)_{due \, to \, image \, well} \tag{6.14}$$

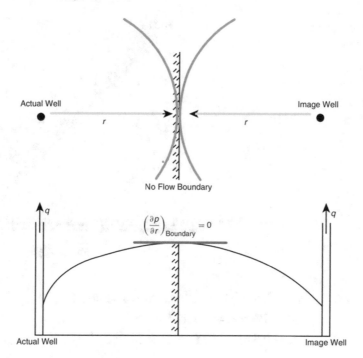

Fig. 6.8 Scheme of images displays a single boundary

Or:

$$(\Delta P)_{total} = \frac{162.6\,q_1 B\mu}{kh}\left[\log\left(\frac{kt}{\emptyset\mu C_t r_w^2}\right) - 3.23 + 0.869\,S\right]$$
$$- \left[\frac{70.6\,q_2 B\mu}{kh}\right] \times E_i\left[\frac{-948\emptyset\mu C_t (2L)^2}{kt}\right] \tag{6.15}$$

The reservoir is considered to be infinite in this equation, with the except for the indicated border. Boundaries usually result in a bigger pressure decrease than predicted for reservoirs with infinite capacity. To simulate the pressure behaviour of a well-positioned inside a range of boundary configurations, the notion of image wells may be expanded.

6.5.1 Example 6.3

A well between two sealing faults is shown in Fig. 6.9 at a distance of 400–600 feet. The well is producing at a steady flow rate of 200 STB per day under a transient flow scenario. Estimate the bottom hole pressure after 10 h. Given the following well data:

Fig. 6.9 Illustrate a well
between two sealing

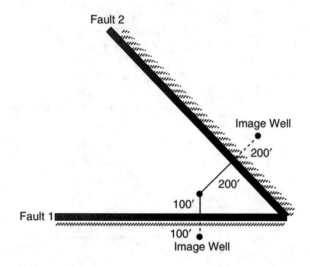

$$pi = 500\,\text{psi},\ k = 600\,\text{md},\ Bo = 1.1\,\text{bbl/STB}$$
$$\varphi = 17\%,\ \mu o = 2.0\,\text{cp},\ h = 25\,\text{ft}$$
$$rw = 0.3\,\text{ft},\ s = 0,\ ct = 25 \times 10^{-6}\,\text{psi}^{-1}$$

Solution

1. Determine the pressure drop caused by the actual well flow rate:

$$\left(p_i - p_{wf}\right) = (\Delta P)_{tactual} = \frac{162.6\,q_1 B\mu}{kh}\left[\log\left(\frac{kt}{\emptyset\mu C_t r_w^2}\right) - 3.23 + 0.869\,S\right]$$

$$\left(p_i - p_{wf}\right) = (\Delta P)_{tactual} = \frac{162.6 \times 200 \times 1.1 \times 2}{60 \times 25}$$

$$\left[\log\left(\frac{60 \times 10}{0.17 \times 2 \times 25 \times 10^{-6} \times (0.3)^2}\right) - 3.23 + 0.869 \times 0\right] = 270\,\text{Psia}$$

2. Determine the extra pressure drop caused by the first fault (image well# 1):

$$(\Delta P)_{image\ well\#\ 1} = P_i - p\,(2L_1, t)$$

$$= -\left[\frac{70.6\,q_2 B\mu}{kh}\right] \times E_i\left[\frac{-9480\mu C_t (2L_1)^2}{kt}\right]$$

$$= -\left[\frac{70.6 \times 200 \times 1.1 \times 2}{60 \times 25}\right]$$

$$\times E_i\left[\frac{-948 \times 0.17 \times 2 \times 25 \times 10^{-6}(2 \times 100)^2}{60 \times 10}\right]21$$

$$[E_i(-0.54)] = 11 \, \text{Psia}$$

3. Determine the effect of the second boundary (image well#2):

$$(\Delta P)_{image \; well\# 2} = p_i - p \, (2L_2, t)$$

$$= - \left[\frac{70.6 \times 200 \times 1.1 \times 2}{60 \times 25} \right] \times E_i \left[\frac{-948 \times 0.17 \times 2 \times 25 \times 10^{-6}(2 \times 200)^2}{60 \times 10} \right]$$

$$= - \; 20.7 \left[E_i(-2.15) \right] = 1.0 \, \text{Psia}$$

4. Determine the total pressure drop:

$$(\Delta P)_{total} = 270 + 11 + 1.0 = 282 \, \text{Psia}$$

5. Determine the bottom hole pressure:

$$P_{wf} = 5000 - 282 = 4718 \; \text{Psia}$$

6.6   Effects of Pressure Changes

Applying the constant pressure-rate scenario also makes advantage of superposition. When there have been two pressure changes, each change will be treated separately using the constant-pressure solution. Accordingly, we must use Eq. 6.15 twice in this specific scenario. To apply the superposition principle to pressure variations in the constant-pressure scenario, the generalized form of Eq. 6.16 will be employed as follows:

$$G_p = \frac{0.111 \emptyset h c r_w^2}{T} \sum_{j=1}^{j=m} \left(\frac{\Delta p_j^2}{z} \right) Q_{pD} \tag{6.16}$$

where G_p = Cumulative gas produced, MSCF, and Q_{pD} = dimensionless total production number for certain boundary conditions (Lee 1982).

For $t_D < 0.01$:

$$Q_{pD} = \left(\frac{t_D}{\pi} \right)^{0.5} \tag{6.17}$$

For $t_D \geq 200$ or

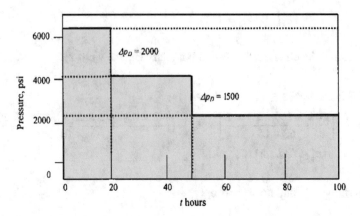

Fig. 6.10 Pressure profile history

$$t_D \alpha Q_{pD} = \frac{-4.29881 + 2.02566 t_D}{ln t_D} \tag{6.18}$$

$$t_D = \frac{0.0002637 kt}{\emptyset \bar{\mu} \, \bar{c} \, r_w^2} \tag{6.19}$$

In terms of pressure-squared treatment

$$\Delta p_j^2 = p_{old}^2 - p_{new}^2 \tag{6.20}$$

And

$$\bar{z} \text{ is calculated at} \left(\frac{p_{old} + p_{new}}{2} \right)$$

As an example, suppose a well has gone through the pressure history depicted in Fig. 6.10.

6.7 Summary

As stated by the superposition principle, once two or more waves overlap in space, the resultant disturbance is equal to the algebraic sum of the individual disturbances. The superposition time function has been applied to interpret transient pressure data recorded under the effect of a fluctuating flow rate. This function is often constructed under the assumption that radial flow equations are valid; however, in reality, there are examples where multiple flow regimes exist. This chapter, addressed the pressure transient test analyses based on an important assumption known as the superposition principle. The superposition constraints also highlighted in details. This chapter covered a variety of topics, included multiple well cases, effects of fluctuating flow rates, reservoir boundary effects, as well as, effects of pressure fluctuations. Examples and solutions were provided for each case.

References

H. C. Slider, Worldwide Practical Petroleum Reservoir Engineering Methods, Penn Well Books, 1983.

John Lee, Well Testing (1982).

Lee, J., Well Testing, Vol. 1, SPE, Textbook Series, Society of Petroleum Engineers of AIME, Dallas, TX, 1982.

Robert Earlougher, Advances in Well Test Analysis (1977).

Roland Horn, Modern Well Test Analysis (1995).

Sabet, M.A. 1991. Well Test Analysis. Gulf publishing company, Houston, Texas. Pp 454.

Slider, H. C. (1976). Practical petroleum reservoir engineering methods. An energy conservation science.

Chapter 7
Well Testing Models

7.1 Reservoir Behaviour

Perez-Rosales (1978), stated that the reservoirs differ in terms of their physical characteristics (type of rock, depth, pressure, size, type of fluid, fluid content, etc.), and there are only a finite number of possible dynamic behaviours of these reservoirs during a well test. This recognition was one of the key components of the integrated methodology. The reason for this is that a reservoir serves as a low-resolution filter, allowing the output signal to only show high contrasts in reservoir properties. Additionally, the combination of three factors that predominate at various points throughout the test (Gringarten et al. 1979; Gringarten 1982, 1985a) yields these dynamic behaviours.

The number of porous media participating in the flow process with varying mobilities (kh/μ) and storativities ($\emptyset C_t h$) is reflected in the fundamental reservoir dynamic behaviour (Grin-Garten 1984, 1985a). Figure 7.1 illustrates these fundamental well test characteristics.

7.2 Dual Porosity Reservoir

In naturally fractured reservoirs, the porosities in the matrix and the fractures are distinct from one another. Even though naturally fractured reservoirs feature irregular fractures, they could be represented by homogeneous dual porosity systems that is equal (Warren and Root 1963).

When compared to how much hydrocarbon is normally kept in the natural fractures, the matrix often holds a significant amount more hydrocarbon. In dual porosity systems, the matrix's permeability is just slightly less than that of natural fractures. As soon as the well begins to produce, fluid effectively and swiftly travels from the high permeability natural fractures to the borehole. Once the natural fissures have formed, the massive quantities of hydrocarbons stored inside the reservoir's matrix

© The Author(s), under exclusive license to Springer Nature Switzerland AG 2023   153
T. A-A. O. Ganat, *Modern Pressure Transient Analysis of Petroleum Reservoirs*,
Petroleum Engineering, https://doi.org/10.1007/978-3-031-28889-0_7

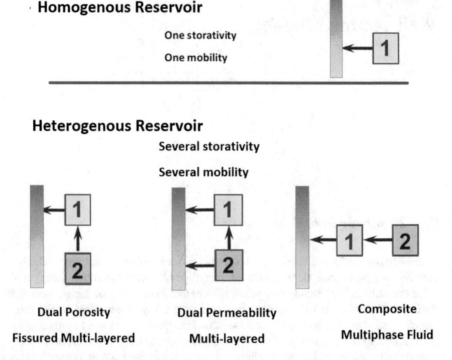

Fig. 7.1 Fundamentals of well test characteristics

begin to flow. These hydrocarbons move to adjacent fractures where almost all of the fluid is transferred to the borehole. Figure 7.2 depicts the characteristic of dual porosity systems as two parallel lines on a semi-log plot, and Fig. 7.3 display Dual porosity systems on a log–log plot.

Early observations of the first semi-log straight line show radial flow as the fluid, which was originally in the fractures, moved toward the borehole. When fluid from the matrix is delivered to the borehole via fractures, a second semi-log straight line is created. The transition between the two semi-log straight lines happens when fluid starts to flow from the matrix to the fractures but has not yet achieved equilibrium. Be mindful that despite the reservoir being naturally fractured, dual porosity, particularly the first semi-log straight line, may not be discernible due to borehole storage effects.

The bulk of the fluid is successfully stored in the secondary porosity system while the primary porosity efficiently controls all flow to the well in a rock that has both primary porosity from the initial deposition and secondary porosity from some other mechanism. Dual-porosity reservoirs include layered reservoirs with distinct differences between strata with high and low permeability, naturally fractured reservoirs, and voluminous carbonates.

According to the double-porosity concepts, the reservoir is made up of blocks of high storativity and low permeability rock matrix that are interconnected to the well

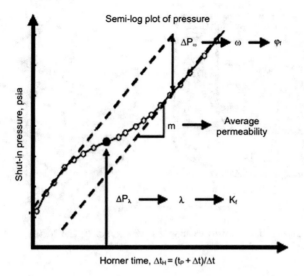

Fig. 7.2 Dual porosity systems on a semi-log plot

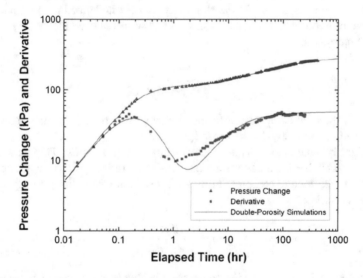

Fig. 7.3 Dual porosity systems on a log–log plot

by low storativity and high permeability natural fissures. M most of the hydrocarbon is stored in the matrix blocks, but because it cannot flow directly to the well, it must first enter the fissure system in order to be produced (Fig. 7.4).

Compared to the homogeneous model, the dual-porosity model has two more variables to characterize it, storativity ratio, ω, and the interporosity flow coefficient, λ.

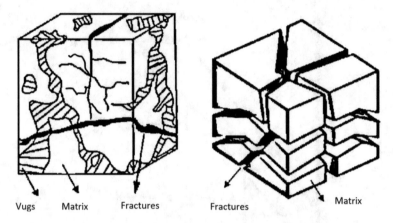

Vugs   Matrix   Fractures   Fractures   Matrix

Fig. 7.4 Type of secondary porosity existing in reservoir rock (Ganat 2020)

Storativity Ratio (ω):

The percentage of the total pore volume connected to one of the porosities in a dual porosity reservoir is known as the storativity ratio. It is defined as: In a naturally fractured reservoir, it refers to the volume of reserves contained inside the fractures and is expressed as:

$$\omega = \frac{\emptyset_f C_f}{\emptyset_f C_f + \emptyset_m C_m} \tag{7.1}$$

where C is the total compressibility and Ø stands for porosity. Fracture and matrix are denoted, respectively, by the subscripts *f* and *m*.

It might be challenging to estimate fracture compressibility. This parameter's value is unavailable for many reservoirs. This leads to the widespread assumption that matrix and fracture compressibility are equivalent. Using this presumption, the equation becomes:

$$\omega = \frac{\emptyset_f}{\emptyset_f + \emptyset_m} \tag{7.2}$$

Therefore, using the following equation, it is simple to get the fracture porosity from the storativity ratio:

$$\emptyset_f = \left(\frac{\omega}{1 - \omega}\right)\emptyset_m \tag{7.3}$$

So, a reasonable indicator of fracture porosity may be provided by well test analysis. Equation 7.2 was used to get the values of the storativity ratio on the presumption that fracture compressibility is equivalent to matrix compressibility. Figure 7.5

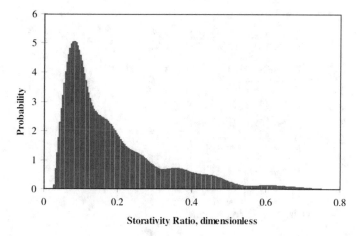

Fig. 7.5 Distribution of the storativity ratio based on actual field data

displays the outcomes. The storativity ratio spans in values from 0.003 to 0.75; the mean value is 1.5, while the median is 0.04.

On a reservoir's capacity to produce in the short term, the storativity ratio has a considerable impact. In dual porosity reservoirs, it is always combined with the interporosity flow coefficient and typically ranges from 0.01 to 0.1.

Interporosity Flow Coefficient (λ):

The ratio of the matrix's (k_m) to the fracture's (k_f) permeability is known as the interporosity flow coefficient in dual porosity reservoirs.

$$\lambda = \sigma r_w^2 \frac{k_m}{k_f} \qquad (7.4)$$

A well test analysis is often used to determine the permeability of the reservoir, k_f. Typically, the borehole radius, r_w, is a well-known value. The shape factor may then be calculated by using the following equation, assuming the matrix permeability value, k_m, is available:

$$\sigma = \frac{\lambda k_f}{r_w^2 k_m} \qquad (7.5)$$

When expressing this number as fracture spacing, simulation models can use it directly. Figure 7.6 illustrates a succession of parallel unit-slope straight lines on a log–log plot of σ versus λ / r_w^2. Each line represents a different k_f / k_m ratio.

A comparable log–log plot may be created when fracture spacing can be considered to be constant in all directions. Figure 7.7 depicts fracture spacing as a function of λ / r_w^2.

where $L_x = L_y = L_z = L_{ma}$.

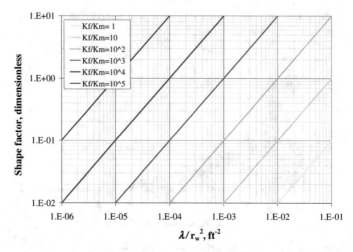

Fig. 7.6 Shape factor versus λ/r_w^2

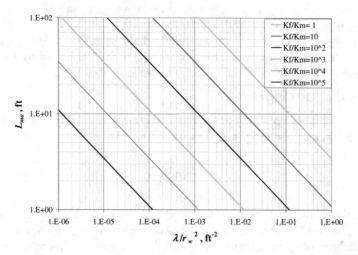

Fig. 7.7 Illustrates fracture spacing as a function of λ/r_w^2

It should be noted that the geometric coefficient explains how the matrix blocks are shaped. In dual porosity reservoirs, the interporosity flow coefficient is always utilized in combination with the storativity ratio and is typically in the range of 10^{-4} to 10^{-8}.

Using field data, a sufficient range of interporosity flow coefficients was determined (Nelson 2001). Figures 7.8 and 7.9 show the matrix and fracture permeability distributions. Although permeability is commonly thought to be log-normally distributed, the data presented show that both matrix and fracture permeability have exponential distributions.

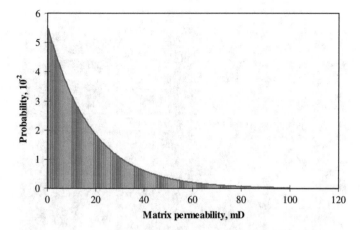

Fig. 7.8 Distribution of matrix permeability in a fracturing reservoir

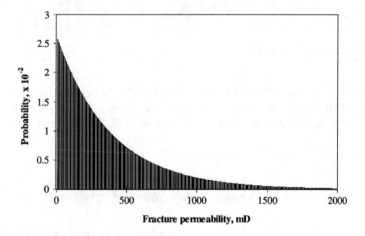

Fig. 7.9 Distribution of fractures permeability in the reservoir

Generally, the fraction of oil or gas that is stored in the fissure system is represented by the storativity ratio, where $\omega = 0.05$ represents 5%. The interporosity flow coefficient describes how readily matrix blocks may enter the fissure system; it is dominated by the matrix/fissures permeability difference, k_m/k_f (Fig. 7.10). Fissure system radial flow, or production from the fissure system with no change in pressure inside the matrix blocks, will be the first flow regime when the well is first placed into production. This initial flow regime normally lasts for a brief time and is frequently concealed by wellbore storage. If not, an Infinite-Acting Radial Flow reaction on the pressure derivative will show up.

When the fissure system begins to produce, a pressure difference is created between the matrix blocks, which are still under initial pressure, and the fissure

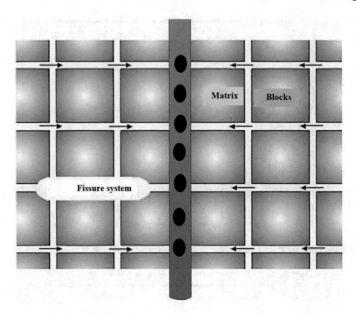

Fig. 7.10 Flow of the fissure system

system, which is under pressure at the wellbore (Fig. 7.11). A transient "dip" in the derivative results from the drawdown temporarily slowing down when the matrix blocks begin to produce into the fissure system, effectively supplying pressure support.

7.2.1 Dual Porosity Pseudo-steady State Interporosity Flow

The most popular dual porosity model implies that the flow between the matrix and the fractures is in a pseudo-steady condition. This situation uses the assumption that there is no pressure drop inside the matrix blocks and that the pressure distribution is uniform. All of the pressure loss occurs as a "discontinuity" at the surface of the blocks, and the pressure response that results from causes a significant "dip" in pressure during the transition (Fig. 7.12).

It could be able to observe the fissure system radial flow in advance if the wellbore storage constant (C) is relatively low. However, the initial flow regime has already been masked with a well storage value of just 0.01 bbl/psi (Fig. 7.13). The dual-porosity transition is detected in the data as soon as storage effects have passed, which might provide a uniqueness problem for the set of data (Figs. 7.14 and 7.15).

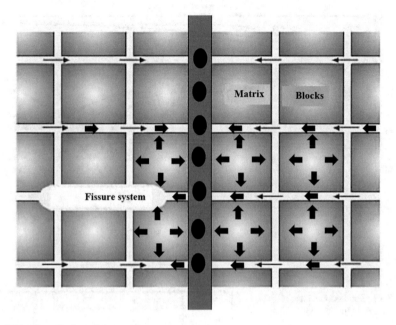

Fig. 7.11 Contribution of the matrix

Fig. 7.12 Dual porosity
system (Pseudo-steady state)

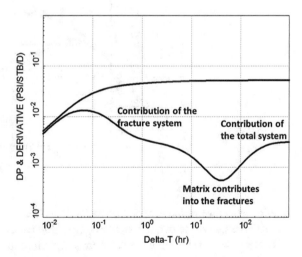

7.2.2 Dual Porosity (Transient Interporosity Flow)

This model presupposes the presence of a pressure gradient and, thus, diffusivity, inside the matrix blocks. Since the geometry of the blocks must be taken into account if the pressure distribution inside the blocks is significant, there are two

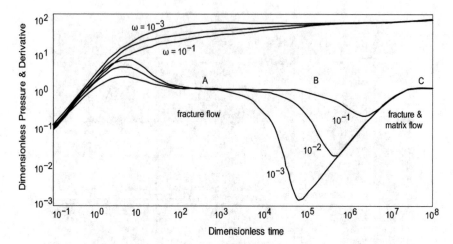

Fig. 7.13 Dual porosity (Pseudo-steady state type-curve)

Fig. 7.14 Show storativity
ratio decrease in direction

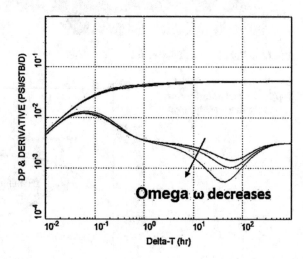

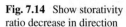

solution systems available, each of which relates to a distinct matrix block geometry. Figure 7.16 shows the two extremely comparable solutions.

In contrast to what we have been contemplating up to this point with the dual-porosity models, the "slab" geometry model suggests rectangular matrix blocks. it is a realistic representation, the "spheres" model offers another straightforward geometry for specifying the boundaries of the mathematical solution. It is challenging to imagine a reservoir made up of spherical matrix blocks, but dual-porosity data sets occasionally fit the "spheres" model better than any other model. This could be because throughout geological time, fluid movements have caused the fracture network to become "vuggy" and the sharp corners of the matrix blocks to become rounded.

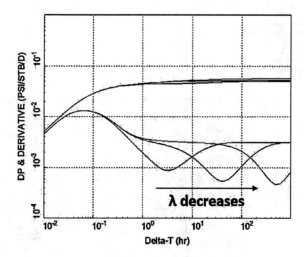

Fig. 7.15 Show interporosity flow coefficient decrease in direction

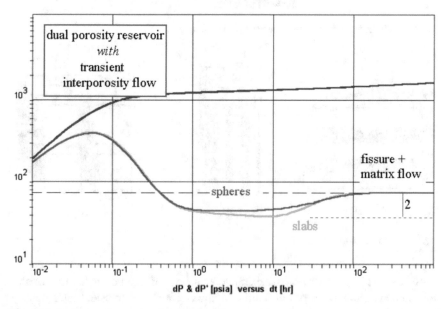

Fig. 7.16 Dual porosity transient interporosity flow

The fissure system radial flow, as depicted in Figs. 7.16 and 7.17, is exceedingly transient and is not seen in practice. Half of the system-wide radial flow value is represented by the semi-log slope/derivative value during the transition.

As can be seen above, this model has a more modest impact on the derivative's form and establishes the moment at which the response changes to total system infinite-acting reservoir fracture.

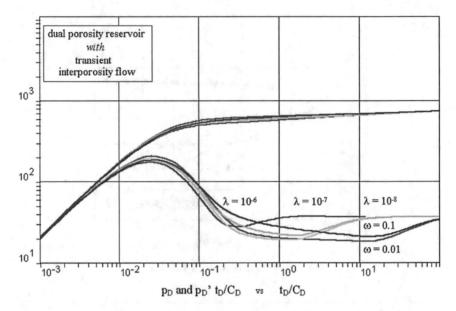

Fig. 7.17 Illustrates transient interporosity flow type-curve

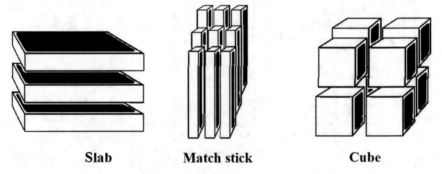

Fig. 7.18 Matrix shapes

As seen in Fig. 7.18, several matrix models can be used to represent fractured reservoir models. The three types of matrix geometry for model transient interporosity flow are slabs, sticks, and cubes.

7.2.3 Example 7.1

Table 7.1 summarizes the data on pressure buildup for a dual-porosity system as provided by Najurieta (1980) and Sabet (1991):

Table 7.1 Buildup data

Δt (h)	Pws (psi)	$\frac{t_p+\Delta t}{\Delta t}$
0.003	6617	3,1000,000
0.017	6632	516,668
0.033	6644	358,334
0.067	6650	129,168
0.133	6654	64,544
0.267	6661	32,293
0.533	6666	16,147
1.067	6669	8074
2.133	6678	4038
4.267	6685	2019
8.533	6697	1010
17.067	6704	506
34.133	6712	253

The following are reservoir and fluid properties data available:

$$P_i = 6789.\,5\,\text{psi},\ \text{pwf} = 6352\,\text{psi at}\ t = 0,$$
$$Qo = 2554\,\text{STB/day},\ Bo = 2.3\,\text{bbl/STB},$$
$$\mu o = 1\,\text{cp},\ t_p = 8611\,\text{hours}$$
$$r_w = 0.\,375\,\text{ft},\ c_t = 8.\,17 \times 10^{-6}\,\text{psi}^{-1},\ \varphi_m = 0.21$$
$$k_m = 0.\,1\,\text{md},\ h_m = 17\,\text{ft}$$

Calculate ω and λ.

Solution

1. Plot p_{ws} versus $(tp + \Delta t)/\Delta t$ on a semi-log scale as illustrated in Fig. 7.19.
2. Figure 7.19 displays two parallel semi-log straight lines with a slope of $m = 32$ psi/cycle.
3. Calculate $(k_t h)$ from the slope m:

$$k_f h = \frac{162.6 q_o B_o \mu_o}{m}$$
$$k_f h = \frac{162.6 \times 25.56 \times 2.3 \times 1.0}{32} = 29{,}848.3\ mD \cdot ft$$

And

$$k_f = \frac{29848.3}{17} = 1756\ \text{mD}$$

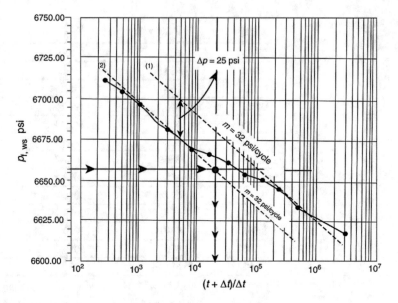

Fig. 7.19 Semi-log plot of the buildup test data (After Sabet 1991)

4. Estimate the vertical distance Δp between the two straight lines:

$$\Delta p = 25 \, \text{psi}$$

5. Determine the storativity ratio, ω:

$$\omega = 10^{-\left(\frac{\Delta p}{m}\right)} = 10^{-(25/32)} = -0.165$$

6. To intersect the two semi-log straight lines, a horizontal line should be drawn across the middle of the transition region. Read the time at the second junction to determine:

$$\left(\frac{t_p + \Delta_t}{\Delta_t}\right)_2 = 20{,}000$$

7. Calculate λ:

$$\lambda = \left[\frac{1}{1-\omega}\right]\left[\frac{(\emptyset h c_t)_m \mu r_w^2}{1.71 k_f t_p}\right]\left(\frac{t_p + \Delta_t}{\Delta_t}\right)_2$$

$$\lambda = \left[\frac{1}{1-0.165}\right]\left[\frac{0.2 \times 17 \times 8.17 \times 10^{-6} \times 1 \times (0.375)^2}{1.71 \times 1756 \times 8611}\right]$$

$$\times 20{,}000 = 3.64 \times 10^{-9}$$

7.3   Dual Permeability Reservoir

The fundamental idea behind the dual permeability form is the same as that of the dual porosity mode, with the addition that matrix cells can also directly link (flow, pressure) to the wells and one another.

Dual-permeability systems presuppose that the matrix and the fracture pore domain, two separate layers but mutually interacting subsystems, may be used to characterize the porous media system. The reservoir is separated into two layers with different permeabilities in the double-permeability idea system, each layer can be perforated. Crossflow is significantly influenced by the pressure difference between the layers (Fig. 7.20).

Another coefficient is included in addition to the storativity ratio and the inter-porosity flow coefficient which is 'K'. This coefficient is the proportion of the first layer's permeability-thickness product to the sum of the two layers

$$k = \frac{k_1 h_1}{k_1 h_1 + k_2 h_2} \tag{7.6}$$

Normally, layer-1 is the layer with the highest permeability, therefore k value will be near 1.0 (Fig. 7.21). Initially, there is no pressure differential between the layers, and the system acts as two homogenous layers without crossflow, in infinite-acting radial flow, with a complete kh of the two layers. Δp develops between the layers as a result of the more permeable layer producing faster than the lower permeable layer, which triggers crossflow to occur. The system eventually behaves as a homogenous reservoir with the combined kh and storativity of the two layers. Figure 7.22 indicates that there is a "concave" in the crossflow double-layer reservoir, which is created when fluids pass through the interlayer from the low permeability layer into the high permeability layer. The "concave" will disappear and curves will overlap when crossflow has been produced over time and pressures in every layer reach equilibrium. Since crossflow decreases flow resistance, the bottom hole pressure in a

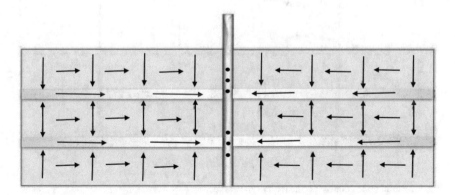

Fig. 7.20 Diagram shows dual-permeability system of a naturally fractured reservoir

crossflow reservoir is lower than it is in a no crossflow reservoir in systematic radial flow, but the bottom hole pressure derivative has the same value of 0.5.

In a dual permeability system, three parameters characterize the heterogeneous dip in the derivative:

Layer Storativity Ratio

$$\omega = \frac{(\emptyset_f C_t h)_1}{(\emptyset_f C_t h)_1 + (\emptyset_f C_t h)_2} \tag{7.7}$$

Interlayer Flow Coefficient

$$\lambda = \sigma r_w^2 \frac{k_2 h_2}{k_1 h_1 + k_2 h_2} \tag{7.8}$$

Permeability contrast

$$k = \frac{k_1 h_1}{k_1 h_1 + k_2 h_2} \tag{7.9}$$

The permeability contrast is influencing also the depth of the dip, where; $K = 1$ (same as Dual porosity-pseudo steady state) and K less than 1 at the shallower dip.

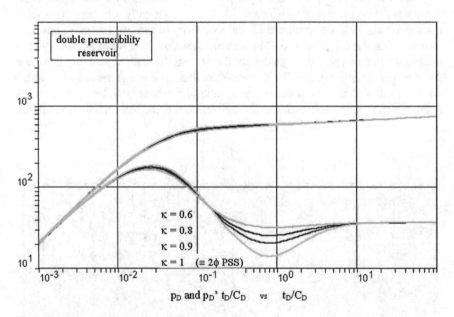

Fig. 7.21 Illustrates double permeability type curve

Fig. 7.22 Shows type curves in the double-layer reservoir with and without crossflow

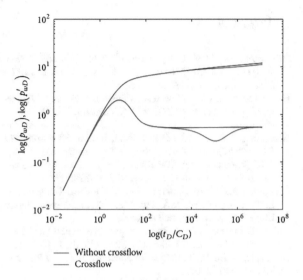

Generally, the transitional dip is controlled by the parameters ω and λ, which have the same impact as in the dual porosity models, which decreases the depth of the dip in comparison to $K = 1$, which results also in the dual-porosity pseudo-steady state solution. As a result, the oil or gas in the low-permeability zone, which corresponds to the matrix blocks, can only be flowed by entering the high-permeability zone, which corresponds to the fissure system, if $k = 1$ and $k_2 h_2 = 0$, respectively. Its behaviour mimics the dual-porosity concept.

7.4 Summary

A well test analysis model is basically a way for simulating the reservoir's hydrocarbon flow process using mathematical or physical methods. The flow in oil or gas strata is physically simulated using a well test model. The two phases of the well-testing study are the identification of the underlying reservoir models and the determination of model-related parameters. The non-uniqueness problem often makes it difficult to choose the best reservoir model when using conventional interpretation approaches. This chapter discusses all well testing models, together with reservoir behaviour, dual porosity reservoirs, and dual permeability reservoirs.

References

Gringarten, A.C. 1984. Interpretation of Tests in Fissured Reservoirs and Multilayered Reservoirs With Double-Porosity Behaviour: Theory and Practice. JPT 36 (4): 549–564. SPE-10044-PA. https://doi.org/10.2118/10044-PA.

Gringarten, A.C. 1982. Flow Test Evaluation of Fractured Reservoirs. In Recent Trends in Hydrogeology, Special Paper 189, ed. T.N. Narasim,237–263. Boulder, Colorado: Geological Society of America.

Gringarten, A.C., Bourdet D.P., Landel, P.A., and Kniazeff, V.J. 1979. A Comparison Between Different Skin and Wellbore Storage Type-curves for Early-Time Transient Analysis. Paper SPE 8205 presented at the SPE Annual Technical Conference and Exhibition, Las Vegas, Nevada, 23–26 September. https://doi.org/10.2118/8205-MS.

Gringarten, A.C. 1985a. Interpretation of Well Test Transient Data. In Developments in Petroleum Engineering—1, ed. R.A. Dawe and D. C. Wilson, 133–196. London and New York City: Elsevier Applied Science Publishers.

Najurieta, Humberto L. "A theory for pressure transient analysis in naturally fractured reservoirs." Journal of Petroleum Technology 32.07 (1980): 1241-1250.

Nelson, R.A.: Geologic Analysis of Naturally Fractured Reservoirs, Gulf Professional Publishing, Woburn, MA, 2001.

Perez-Rosales, C. 1978. Use of Pressure Build up Tests for Describing Heterogeneous Reservoirs. Paper SPE 7451 presented at the SPE Annual Fall Technical Conference and Exhibition, Houston, 1–3 October. https://doi.org/10.2118/7451-MS.

Sabet, M.A. 1991. Well Test Analysis. Gulf publishing company, Houston, Texas. Pp 454.

Tarek Ganat. (2020). Fundamentals of Reservoir Rock Properties. Publisher Springer. https://doi.org/10.1007/978-3-030-28140-3_1

Warren, J.E., Root, P.J.: The behaviour of naturally fractured reservoirs. SPE J. 3(3), 245–255 (1963).

Chapter 8
Gas Well Testing

8.1 Gas Flow Behaviour

Within the range of the given pressures, modeling liquid flow for well test interpretation includes both constant values of density and compressibility factor into account. This presumption is invalid in the case of gas flow when the gas compressibility factor is also taken into consideration for the better mathematical model. Normal linearization of the gas flow model allows the liquid diffusivity solution to fulfil the gas flow characteristics. Three treatments are taken into consideration for the linearization, depending on the viscosity-compressibility product: square of pressure squared, pseudopressure, or linear pressure. Drawdown tests are best analyzed using the pseudopressure function when borehole storage conditions are not significant. Additionally, pseudotime best reflects the thermodynamics of gases because the viscosity-compressibility product is very sensitive to gas flow. To linearize pseudotime and pseudopressure, for instance, buildup pressure tests must be performed. For interpreting well test results, the traditional straight-line approach has been utilized often. Its weaknesses include its inability to provide verification and its accuracy in identifying the beginning and end of a certain flow regime. Compressibility for gas is inversely correlated with pressure; at first, it is high compression is easy at first but gets progressively more difficult. As molecules are driven closer together and collide more often under pressure, gas viscosity rises (Fig. 8.1).

T. A-A. O. Ganat, *Modern Pressure Transient Analysis of Petroleum Reservoirs*, Petroleum Engineering, https://doi.org/10.1007/978-3-031-28889-0_8

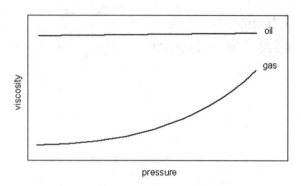

Fig. 8.1 Show the viscosity is inversely proportional to the pressure

8.2 Diffusitivity Equation for Gas Flow

Three concepts were used to derive the diffusivity equation, conservation of mass, the equation of state for slightly compressible liquids, and Darcy's law. This equation is linear, which makes it much simpler to find solutions (like the one for the Ei-function) and enables us to apply superposition in time and space to create solutions from simple single-well solutions for complicated flow geometries and variable rate records.

$$\frac{1}{r}\left(r\frac{\partial p}{\partial r}\right) = \frac{\emptyset \mu C_t}{0.0002637\,k}\frac{\partial p}{\partial t} \tag{8.1}$$

8.2.1 Pseudopressure Function

The pseudopressure method is the most popular of the conventional analytical techniques used to analyze gas wells. Al-Hussainy et al. (1966) proposed the concept of pseudopressure. The meaning of pseudopressure is seen in Eq. (8.2):

$$P_p(p) = 2\int_{P_o}^{p}\frac{p}{\mu z}d_p \tag{8.2}$$

where p_o is the reference pressure. In this work, p_o was specified as the initial reservoir pressure. The diffusivity equation takes the final following form:

$$\frac{1}{r}\frac{\partial}{\partial r}\left(r\frac{\partial p_p}{\partial r}\right) = \frac{\emptyset \mu C_t}{0.0002637\,k}\frac{\partial p_p}{\partial t} \tag{8.3}$$

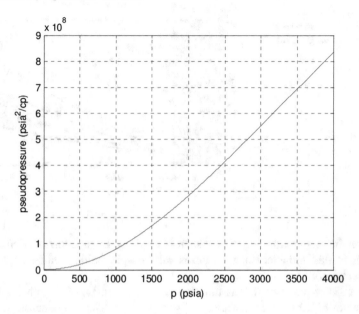

Fig. 8.2 Shows pseudopressure versus real pressure

When pressure is substituted for pseudopressure in Eq. 8.3, the equation takes on the same form as the diffusivity equation for slightly compressible liquids. As a result of the fact that the product μc_t strongly depends on pressure, this solution is nonlinear. Fortunately, studies have demonstrated that the equation can be treated as linear and that the Ei-function is valid for gases if μc_t is assessed at the pressure at the start of a flow period until the point at which boundaries start to significantly affect the pressure drop at the well; in other words, as long as the reservoir is infinite-acting.

Pseudopressure, p_p, is solely dependent on the relationship between pressure and μz. When the gas's PVT table is provided, the $p_p - p$ relation is always known. It is important to understand that is not time-dependent. Figure 8.2 depicts an illustration of a $p_p - p$ curve.

8.2.1.1 Pressure-Squared and Pressure Approximations

Pseudopressure is obtained from Eq. 8.3 by assuming that the product z is constant.

$$p_p(p) = \frac{1}{\mu z}\left(p^2 - p_o^2\right) \tag{8.4}$$

So that the diffusivity equation is:

$$\frac{1}{r}\frac{\partial}{\partial r}\left(r\frac{\partial p^2}{\partial r}\right) = \frac{\emptyset \mu C_t}{0.0002637\,k}\frac{\partial p^2}{\partial t} \tag{8.5}$$

Fig. 8.3 μz versus pressures at a variety of gas gravities

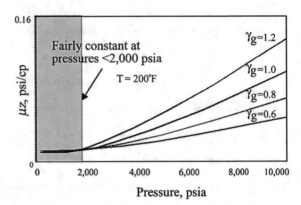

When the assumption that z is constant is valid, the Ei-function solution is valid in terms of this variable. Equation 8.5 is valid despite being nonlinear (pressure-dependent μc_t).

The validity range of this assumption is depicted in Fig. 8.3 with a reservoir temperature of 200 °F and a variety of gas gravities. The z product is relatively constant at pressures less than 2000 psia (the shaded area in Fig. 8.3). The outcomes are consistent at temperatures ranging from 100 to 300 °F.

Pseudopressure is generated from Eq. 8.4 by assuming that the group p/z is constant.

$$p_p(p) = \frac{p}{\mu z}(p - p_o) \tag{8.6}$$

So that the diffusivity equation is:

$$\frac{1}{r}\frac{\partial}{\partial r}\left(r\frac{\partial p}{\partial r}\right) = \frac{\emptyset \mu C_t}{0.0002637k}\frac{\partial p}{\partial t} \tag{8.7}$$

The independent variable is now p, and the Ei-function is valid for pressure when the presumption that p/z is constant is valid. This is true although Eq. 8.7 is nonlinear, however, it only applies to infinite-acting reservoirs. For a reservoir temperature of 200 °F and a range of various gas gravities, Fig. 8.3 illustrates the range of validity of this assumption (the shaded region in Fig. 8.4). At pressures above 3000 psia and at various temperatures between 100 and 300 °F, the group p/z remains constant.

These findings imply that the selection of a variable for gas well-flow equations relies on the circumstances. The pressure-squared approximation is only applicable to low pressures ($P > 2000$ psia), the pressure approximation is only applicable to high pressures ($P > 3000$ psia), and the pseudopressure transformation is only applicable to all pressure ranges. The pseudopressure is nearly always the best variable to utilize when analyzing pressure transient tests using the software. Only pressure or pressure-squared techniques are workable for hand analysis.

Fig. 8.4 $p/\mu z$ versus
pressure at a variety of gas
gravities

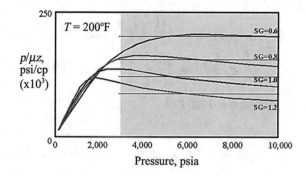

Normalized Pseudopressure:

When given pressure dimensions, the pressure function can be normalized:

$$p_{pn} = p_i + \frac{\mu_i z_i}{p_i} \int_{p_1}^{p_2} \frac{p}{\mu_p z_p} dp \qquad (8.8)$$

The equations for oil solutions may be employed thanks to the normalized pseudopressure, normalized regard to gas characteristics under static circumstances, p_i.

8.2.1.2   Example 8.1

The gas properties as functions pressure (Gas gravity is 0.7 and $T = 2000$ F) are given in Table 8.1. Determine Gas Pseudopressure.

Solution
Use the trapezoidal rule for numerical integration. For $p = 150$ psia,

$$p_p(p) = 2 \int_{p_o}^{p} \frac{p}{\mu z} dp$$

Table 8.1 Gas properties

Pressure P (psia)	Gas viscosity (cP)	Compressibility factor 1/psia	$p/\mu z$ (psia/cP)
150	0.01238	0.9856	12,290
300	0.01254	0.9717	24,620
450	0.01274	0.9582	36,860

$$p_p(150) = 2\frac{\left[\left(\frac{p}{\mu z}\right)_o + \left(\frac{p}{\mu z}\right)_{150}\right]}{2} \times (150 - 0)$$

$$p_p(150) = 2\frac{[0 + 12.290]}{2} \times (150) = 1.8 \times 10^6 \text{ Psia}^2/\text{cP}$$

For pressure $= 3000$ Psia

$$p_p(300) = 1.8 \times 10^6 + 2 \times \frac{\left[\left(\frac{p}{\mu z}\right)_{150} + \left(\frac{p}{\mu z}\right)_{300}\right]}{2} \times (300 - 150)$$

$$p_p(300) = 1.8 \times 10^6 + 2 \times \frac{[12.290 + 24.620]}{2} \times (300 - 150)$$

$$= 7.4 \times 10^6 \text{ Psia}^2/\text{cP}$$

8.2.2   Pseudotime Function

Pseudo-time is a mathematical time function that takes into consideration the variable total (formation) porosity ($\emptyset$) with respect to time and pressure, together with the variable compressibility (c_t) and viscosity (μ_g) of gas.

The gas is very compressible, and its compressibility (c), which varies with pressure, is not constant. A similar issue exists with gas viscosity (μ_g), although not to the same extent (Fig. 8.5).

The formula for gas flow in a reservoir is quite similar to the formula for liquid flow. Analytical equations are solved in well testing after being predicated on specific assumptions. So, if pressure and time are substituted with real and pseudo-pressure and pseudo-time, the liquid flow solution may be utilized for analysis and forecasting of gas well tests:

$$\text{Pseudo} - \text{Pressure} = \text{Constant}^*\log(\text{Pseudo} - \text{time}) + \ldots$$

Fig. 8.5 Pressures versus gas compressibility

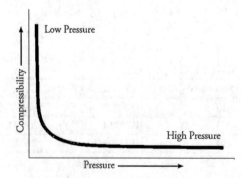

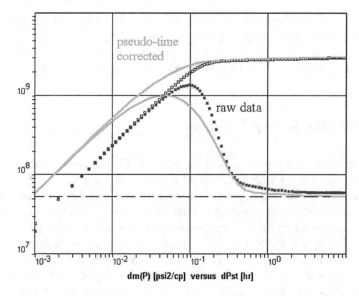

Fig. 8.6 Gas buildup test—pseudo-time corrected

The assumption that C_t is constant causes an error in the early time of the log–log and derivative data in a tight reservoir with significant drawdowns. Using the gas pseudo-time function created by Agarwal in (1979), the raw data is replotted (Fig. 8.6, solid green line).

$$t(p, t) = \int_{t_o p_o}^{t,p} \frac{\partial t}{\mu(p)c(p)} \qquad (8.9)$$

8.2.2.1   Normalized Pseudotime

To enable the use of the conventional superposition calculations, the pseudo-time can also be normalized, as recommended by Meunier et al. (1987), in such a way that at the late time the normalized function t_{pn} becomes comparable to Δt:

$$t_{pn} = (\mu C_t)_i \int_{o}^{\Delta t} \frac{\partial t}{\mu(p)C_t(p)} \qquad (8.10)$$

The new variables are normalized, or multiplied by the relevant constants, to give them the same units and ranges as pressure and time, respectively. With these adjustments, it is possible to derive the equations for the analysis of gas wells by straightforward substitution from the formulas for the analysis of oil well testing in terms of normalized pseudopressure and pseudotime, also known as adjusted pressure and adjusted time. Of course, a computer is needed for the conversions, these modifications are offered by commercial well-test analysis tools.

8.3 Gas Well Testing Techniques

For gas wells, several test techniques for deliverability have been established. The well is produced at a variety of stabilized flow rates during flow-after-flow testing, and the stabilized bottom hole pressure is recorded after each flow. Without an intermediate shut-in time, each flow rate is set in succession. When the bottom hole flowing pressure is stabilized, a single-point test is carried out by flowing the well at a single rate. The drawback of the flow-after-flow test's extensive testing timeframes needed to attain stability at each rate led to the development of this kind of test.

Isochronal and modified isochronal tests were developed to shorten test durations for wells with slow stabilization times. It is common practice to alternate producing at a continuously decreasing sand-face rate without pressure stabilization. This conducted with shutting-in and allowing the well to reach the average reservoir pressure before starting the next flow period to conduct a series of single-point tests known as isochronal tests. When performing the modified isochronal test, the same techniques are utilized, but the flow and shut-in durations are both the same length (but not necessarily the same as the flow periods).

8.4 Flow After Flow Test

A flow after flow test, also known as a backpressure test or a four-point test, includes producing the well at a range of stabilized flow rates and monitoring the stabilized bottom hole flow pressure at the sand-face. With or without a short shut-in phase, different flow rates are sequentially produced. Standard flow-after-flow tests are usually conducted with a rising flow rate sequence, and if steady flow rates are achieved, the rate sequence has minimal influence on the test.

The requirement to sustain the shut-in and flowing phases until stabilization, especially in low-permeability formations where it takes a while to reach stable flowing conditions, is a significant drawback of the flow-after-flow test. Figure 8.7 shows a flow-after-flow test.

Both the exponential and the quasi-quadratic backpressure relationships can be used to evaluate flow-after-flow testing. In both situations, it is presummated that the variables in these formulas remain constant across the whole range of pressures. The explanation using the exponential relation is as follows. We plot every flow rate separately.

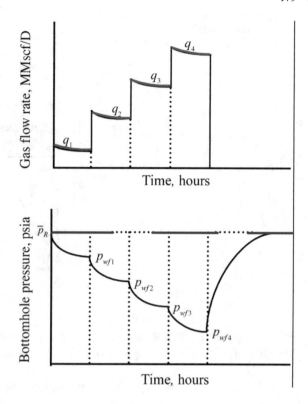

Fig. 8.7 Display flow after flow test

8.4.1 *Rawlins-Schellhardt Analysis Method*

The relationship between gas flow rate and pressure may be described as follows, according to Rawlins and Schellhardt's (1936) postulation, which was based on the examination of flow data collected from a large number of gas wells:

$$q_g = c\left(\overline{p_r}^2 - p_{wf}^2\right)^n \tag{8.11}$$

where:

q_g is gas flow rate, Mscf/day, $\overline{p_r}$ is average reservoir pressure, psi, n is the exponent and C is performance coefficient, Mscf/day/psi^2.

The purpose of the exponent n is to take into account the extra pressure drop brought on by the turbulence in the high-velocity gas flow. The exponent n can range from 1.0 for a perfectly laminar flow to 0.5 for a completely turbulent flow, depending on the flow circumstances. Equation 8.11 includes the performance coefficient C to take reservoir flow geometry, fluid characteristics, and rock parameters into consideration.

Equation 8.11 is often known as the deliverability equation or the back-pressure equation. It is possible to calculate the gas flow rate q_g at any bottom-hole flow

Fig. 8.8 Well deliverability
graph

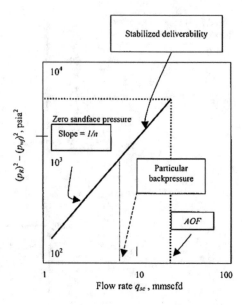

pressure and create the IPR curve if the equation's coefficients (*n*) and (*C*) can be
identified. The logarithm of Eq. 8.11 of both sides yields:

$$\log(q_g) = \log(c) + n \log(\overline{p_r}^2 - p_{wf}^2) \tag{8.12}$$

According to the flowing expression, the deliverability exponent n may be
calculated from any two points on the straight line, i.e. $(q_{g1}, \Delta p_1^2)$ and $(q_{g2}, \Delta p_2^2)$.

$$n = \frac{\log(q_1) - \log(q_2)}{\log(\Delta p_1^2) - \log(\Delta p_2^2)} \tag{8.13}$$

Equation 8.13 predicts that a plot of q_g versus $\left(\overline{p_r}^2 - p_{wf}^2\right)$ on log–log scales
should result in a straight line with an *n*-slope. By graphing $\left(\overline{p_r}^2 - p_{wf}^2\right)$ versus q_g
on the logarithmic scales, a straight line with a slope of (1/*n*) is created in the natural
gas sector when the plot is conventionally inverted. The deliverability graph, often
known as the back-pressure plot, is a plot that is schematically depicted in Fig. 8.8,
(Ahmed 2019).

The performance coefficient *C* may be calculated from any point on the straight
line given *n*:

$$c = \frac{q_g}{\left(\overline{p_r}^2 - p_{wf}^2\right)^n} \tag{8.14}$$

The usual method for determining the coefficients of the back-pressure equation or any other empirical equation is by analyzing the findings of gas well testing.

8.4.2 Example 8.2

A gas well located in a low-pressure reservoir was subjected to a flow-after-flow test. Find the n and C values for the deliverability equation, AOF, and flow rate for P_{wf} = 175 psia based on the test results (Tables 8.2 and 8.3).

Solution
Construct $\left(\overline{p_r}^2 - p_{wf}^2\right)$ versus Q_g on log–log paper as shown in Fig. 8.9 and determine the exponent n.

It is clear from the plot that tests points 1 and 4 lie on a straight line which are used to calculate n:

$$n = \frac{\log(q_1) - \log(q_4)}{\log(\Delta p_1^2) - \log(\Delta p_4^2)}$$

$$n = \frac{\log(2730) - \log(5550)}{\log(1.985 \times 10^{-3}) - \log(4.301 \times 10^{-3})} = 0.92$$

From test point 4, calculate C:

Table 8.2 Flow after flow test data

q_o MSCF / day	P_{wf} psi
0	201
2730	196
3970	195
4440	193
5550	190

Table 8.3 Function plotting data

Q_g MSCF/ day	P_{wf} psi	$\left(\overline{p_r}^2 - p_{wf}^2\right) \times 10^{-3}$, psi²
0	201	40.4
2730	196	1.985
3970	195	2.376
4440	193	3.152
5550	190	4.301

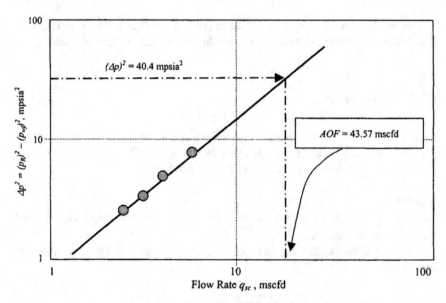

Fig. 8.9 $(\overline{p_r}^2 - p_{wf}^2)$ versus Q_g

$$c = \frac{q_g}{\left(\overline{p_r}^2 - p_{wf}^2\right)^n}$$

$$c = \frac{5550}{\left(4.301 \times 10^{-3}\right)^{0.92}} \quad 2.52 \, \text{MScf/Psia}$$

So, the deliverability equation is:

$$q_g = 2.52\left(\overline{p_r}^2 - p_{wf}^2\right)^{0.92}$$

At Bottom hole pressure, $P_{wf} = 0$ Psia:

$$q_g(AOF) = 2.52 \, \left((201)^2 - (0)^2\right)^{0.92}$$
$$q_g(AOF) = 43579 \, \text{MScf/d}$$

At Bottom hole pressure, $P_{wf} = 175$ Psia:

$$q_g(AOF) = 2.52 \, \left((201)^2 - (175)^2\right)^{0.92}$$
$$q_g = 11812.691 \, \text{MScf/d}$$

8.5 Isochronal Test

An isochronal or equal time test is preferable if the time required for the well to stabilize on each choke size or production rate is excessive.

To conduct the isochronal test, the well is alternately produced, shut in, and allowed time to achieve the average reservoir pressure before the next production phase begins. Each flow cycle's pressures are measured throughout a number of time periods. The pressures should be measured at intervals that correspond to the start of each flow phase. Because the isochronal test builds to beginning pressure more quickly after short flow periods than a flow-after-flow test does, it is more practical for low-permeability formations. A final stabilized flow point is established at the end of the test. Isochronal test is shown in Fig. 8.10 (Cullender 1955).

Plotting the values of $\left(\bar{p_r}^2 - p_{wf}^2\right)$ computed at the particular time intervals versus qo yields n, which is defined by the slope of the line. One test must be stable to calculate a value for C. Figure 8.10 depicts the optimal behaviour of generating rate and pressure as a function of time.

The procedures for conducting an isochronal test are as follows:

1. After shutting down the well, open it at a steady production rate and measure the flow during designated periods. The overall production time at each rate may be shorter than the stabilizing time.
2. Close the well and let the pressure rise to the initial reservoir pressure.
3. Open the well at a different production rate and collect pressure readings at regular intervals.
4. Close the well until the bottom hole pressure equals reservoir pressure.
5. Repeat these steps for various rates.

Fig. 8.10 Typical steps of the isochronal test

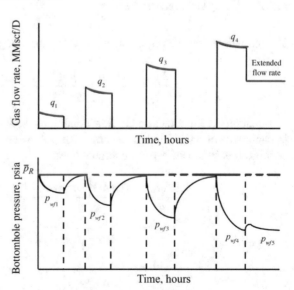

8.5.1 Rawlins and Schellhardt's Analysis

Rawlins and Schellhardt's analysis empirical solution for analyzing data from flow-after-flow test results is given below in log form:

$$\log(q) = \log(C) + n\left\{\log\left[p_p(\overline{p}) - p_p(p_{wf})\right]\right\} \tag{8.15}$$

Plot transient data for isochronal tests that have been collected at the same time intervals relative to the beginning of each flow cycle but measured at various flow rates. Due to the parallelism of the lines connecting data points with the same fixed flow time, the value of n is constant and time-independent. For each isochronal line, a distinct intercept should be determined since the intercept, log (C), is a function of time. This "transient" intercept is log (C_t):

$$\log(q) = \log(C_t) + n\left\{\log\left[p_p(\overline{p}) - p_p(p_{wf})\right]\right\} \tag{8.16}$$

Initial reservoir pressure is replaced by p_s in the modified equation. Where, $p_p =$ pseudopressure, psia^2/cp, and $p_s =$ stabilized shut-in BHP measured just before the start of a deliverability test, psia.

The Rawlins-Schellhardt approach of the isochronal test is to plot $\log\left[\Delta p_p = p_p(p_s) - p_p(p_{wf})\right]$ versus $\log(q)$ for each time, giving a straight line of slope $1/n$ and an intercept of $\left\{-\frac{1}{n}\left[\log(C_t)\right]\right\}$.

8.5.2 Houpeurt Analysis

The following Houpeurt equation are used for analyzing flow-after-flow tests:

$$\frac{\Delta p_p}{q} = \frac{p_p(\overline{p}) - p_p(p_{wf})}{q} = a + bq \tag{8.17}$$

Equation 8.17 assumes stable flow conditions, however, observed transient data is captured during isochronal testing. The following formula for transient flow conditions is hence for every isochronal line:

$$\frac{\Delta p_p}{q} = \frac{p_p(p_s) - p_p(p_{wf})}{q} = a_t + bq \tag{8.18}$$

$$a_t = \sqrt{L_f^2 + b_f^2} \tag{8.19}$$

$$b_f = 0.02878\left[\frac{kt}{\emptyset\mu C_t}\right]^{1/2} \tag{8.20}$$

where

$$a_t = \frac{1.422 \times 10^6 T}{k_g h} \left[\ln(\frac{r_d}{r_w}) - \frac{3}{4} + S \right]$$ (8.21)

$$b = \frac{1.422 \times 10^6 DT}{k_g h}$$ (8.22)

where, Δp_p = pseudopressure change since the start of the test, psia^2/cp, a_t is transient deliverability coefficient, psia^2–cp/MMscf-D, L_f = fracture half-length, ft, b_f depth of investigation of along minor axis in fractured well, ft, k_g = permeability to gas, mD, T = reservoir temperature, °R, and D = non-Darcy flow constant, r_d = effective drainage radius, ft, D/Mscf, and h = net formation thickness, ft.

According to the formulation of Eq. 8.18, a plot of $p_p/q = [p_p(p_s) - pp(p_{wf},s)]/q$ versus q should result in a straight line with slope b and intercept a. The stabilized point may then be included in this concept, and using the stabilized point's coordinates, a stabilized intercept, a, can be calculated. The slope b does not change.

8.5.3 Example 8.3

Calculate the well's absolute open flow (AOF) using the Rawlins and Schellhardt analyses as well as Houpeurt analyses. Table 8.4 lists the results of the isochronal test. Assuming pb is 14.65 psia.

Solution

1. Using Rawlins-Schellhardt analysis method.

Plot $\Delta p_p = p_p(p_s) - p_p(p_{wf})$ versus q on the log–log chart (Fig. 8.11) along with the extended flow test point. Table 8.5 gives the plotting functions.

Using least-squares regression analysis, determine the deliverability exponent, n, for every line or isochron. Observe that the first data point for every isochron is disregarded in the calculations that follow because it does not align with the data points for the last three flow rates (Fig. 8.11). Table 8.6 displays the deliverability exponents determined using least-squares regression analysis for every isochron. The average number of the n values is 0.89 in arithmetic terms.

As $0.5 \leq \bar{n} \leq 1.0$, AOF can be estimated graphically using Fig. 8.12. In this example, the AOF will be computed. First, using the coordinates of the stabilized, extended flow point and $n = \bar{n}$.

$$c = \frac{q_s}{\left[p_p(p_s) - p_p(p_{wf,s}) \right]^n}$$

$$c = \frac{1.156}{\left[2.443 \times 10^6 \right]^{0.89}} = 2.4 \times 10^{-6}$$

Table 8.4 Isochronal test data

Time (hrs)	q (MMScf/d)	$P_{wf\,(Psia)}$	$P_P(Psia^2/Cp)$
0.5	0.983	344.7	9.639×10^6
1	0.977	342.4	9.541×10^6
2	0.970	339.5	9.417×10^6
3	0.965	337.6	9.338×10^6
0.5	2.631	329.5	9.003×10^6
1	2.588	322.9	8.735×10^6
2	2.533	315.4	8.437×10^6
3	2.500	310.5	8.246×10^6
0.5	3.654	318.7	8.567×10^6
1	3.565	309.5	8.207×10^6
2	3.453	298.6	7.792×10^6
3	3.390	231.9	7.544×10^6
0.5	4.782	305.5	8.053×10^6
1	4.625	293.6	7.614×10^6
2	4.438	279.6	7.099×10^6
3	4.318	270.5	6.779×10^6
214	*1.156*	*291.6*	7.529×10^6

Extended flow test

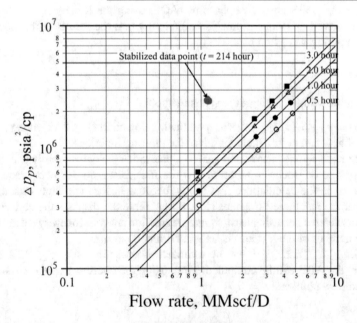

Fig. 8.11   Isochronal test plot using Rawlins-Schellhardt analysis

Table 8.5 Plotting function data

Time (hrs)	q (MMScf/d)	$\Delta p_p \mathrm{Psia}^2/$ (Cp)	Time (h)	q (MMScf/d)	Δp_p (Psia^2/ Cp)
0.5	0.983	0.333×10^6	2.0	3.654	1.404×10^6
	0.977	0.431×10^6		3.565	1.764×10^6
	0.970	0.554×10^6		3.453	2.179×10^6
	0.965	0.633×10^6		3.390	2.428×10^6
1.0	2.631	0.969×10^6	3.0	4.782	1.918×10^6
	2.533	1.236×10^6		4.625	2.358×10^6
	2.500	1.726×10^6		4.438	2.873×10^6
				4.318	3.192×10^6
			214	1.156	2.44×10^6

Table 8.6 Deliverability exponents

Time (hrs)	Deliverability exponents (n)
0.5	0.88
1.0	0.91
2.0	0.89
3.0	0.88

So, the calculated AOF potential:

$$q_{AOF} = C\left[p_p(\overline{p}) - p_p(p_b)\right]^n$$
$$q_{AOF} = 2.4 \times 10^{-6}\left[9.9715 \times 10^6 - 2098.7\right]^{0.89}$$
$$= 4.04\,\mathrm{MMScf/d}$$

To estimate the AOF from the graph, then calculate the pseudopressure first at p_b:

$$\Delta p_p = p_p(p_s) - p_p(p_b) = 9.9715 \times 10^6 - 2098.7 = 9.969 \times 10^6$$

Draw a line of slope 1/n over the stabilized extended flow point and extrapolate the line to the flow rate at $\Delta p_p = p_p(p_s) - p_p(p_b)$, and read the q_{AOF}. From the graph, the $q_{AOF} = 4.04$ MMscf/D.

1. Using the Houpeurt analysis method

Plot on a Cartesian graph, plot $\Delta p_p/q = [p_p(p_s) - p_p(p_{wf})]/q$ versus q (Fig. 8.13). The plotting functions are existing in Table 8.7. For each time, draw the best-fit lines across the isochronal data sets. Because the point related to the lowest rate fits on the same straight line for every flow time, all four data sets will be utilized for the evaluation of every isochron.

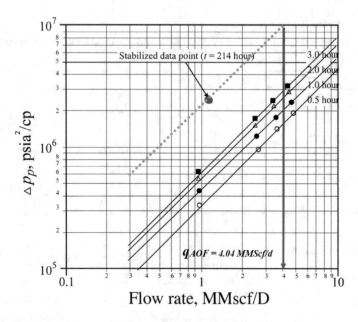

Fig. 8.12 Estimating Q_{AOF} graphically

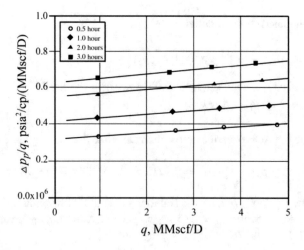

Fig. 8.13 Houpeurt analysis of isochronal test data

Estimate each line's or isochron's slope b. Table 8.8 summarizes the b values derived from least-squares regression analysis. The arithmetic average of b value is 2.074×10^4 Psia²/cp/ /(MMscf/D) ².

Table 8.7 Plotting function data

Time (h)	q (MMScf/d)	$\Delta p_p/q/q$ (Psia^2/ MMScf/d)	Time (h)	q (MMScf/d)	$\Delta p_p/q/q$ (Psia^2/MMScf/d)
0.5	0.983	0.387×10^5	2.0	0.970	5.707×10^5
	2.631	3.682×10^5		2.533	6.058×10^5
	3.654	3.843×10^5		3.453	6.311×10^5
	4.782	4.011×10^5		4.438	6.473×10^5
1.0	0.977	4.410×10^5	3.0	0.965	6.564×10^5
	2.588	4.777×10^5		2.500	6.903×10^5
	3.565	4.949×10^5		3.390	7.162×10^5
	4.625	5.098×10^5		4.318	7.392×10^5

Table 8.8 Slop "b" for the modified isochronal test of Houpeurt analysis

Time (h)	Psia^2/cp/(MMscf/D)^2 (b)
0.5	1.644×10^4
1.0	1.904×10^4
1.5	2.255×10^4
2.0	2.492×10^4

Estimate the deliverability line intercept using $\Delta pp/q = 2.113 \times 10^6$ psia^2/cp/(MMscf/d) at the stabilized point.

$$a = (\Delta p_p/q) - bq$$
$$a = (2.113 \times 10^6) - (2.074 \times 10^4 \times 1.156)$$
$$= 2.109 \times 10^6 \, \text{psia2/cp/(MMscf/d)}$$

Determine the q_{AOF} using the average slope b along with the stabilized value of a.

$$q_{AOF} = \frac{-a + \sqrt{a^2 + 4b[p_p(p_s) - p_p(p_b)]}}{2b}$$

$$q_{AOF} = \frac{-2.109 \times 10^6 + \sqrt{(2.109 \times 10^6)^2 + 4(2.074 \times 10^4)(9.97 \times 10^6)}}{2(2.074 \times 10^4)}$$

$$= 4.53 \, \text{MMScf/d}$$

8.6   Modified Isochronal Test

Even after short flow times, the time required to buildup to the average reservoir pressure before flowing for a particular time may be difficult. As a result, a modification to the isochronal test (Katz et al. 1959) was proposed to minimize test times. The modified isochronal test aims to collect the same test data as an isochronal test.

Similar to the isochronal test, two lines are obtained: one for the isochronal data and one via stabilized point. The ideal stable deliverability curve is represented by the last line. Although the modified isochronal test technique does not provide a perfect isochronal curve, it comes very close. Figures 8.14 and 8.15 show the pressure and flow rate order for the modified isochronal test.

8.6.1   Example 8.4

Applying the Rawlins and Schellhardt and Houpeurt analytical methodologies determines the AOF using the data from Well-R3 as shown below. The test data are shown in Table 8.9.

$p_p(p_b) = 5.093 \times 10^7$ psia2/cp $p_b = 14.65$ psia,
$h = 6$ feet, $r_w = 0.1875$ feet,
$\emptyset = 0.2714$, $T = 540$ °R (80 °F),
$\overline{\mu_g} = 0.015$ cp, $\overline{z} = 0.97$,
$\overline{c_g} = 1.5103$ Psia^{-1}, $\gamma_g = 0.75$,
$S_w = 0.30$, $c_f = 3106$ Psia^{-1},

$A = 640$ acres (centred gas well in a square drainage area).

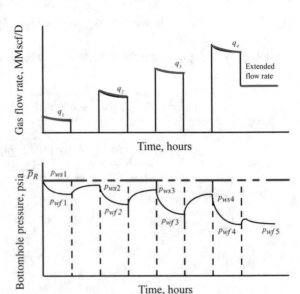

Fig. 8.14 Modified Isochronal test stages

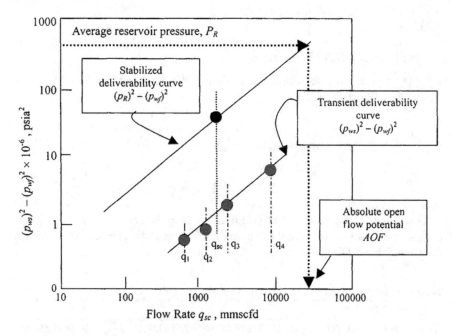

Fig. 8.15 Modified isochronal test data points

Table 8.9 Modified isochronal test data

P_{wf} (Psia)				
Time (h)	q = 1.520 MMScf/d	q = 2.041 MMScf/d	q = 2.688 MMScf/d	q = 3.122 MMScf/d
0 (P_s)	706.6	706.6	703.5	701.2
0.5	655.6	724.5	578.5	541.7
1.0	653.6	620.7	573.9	537.8
1.5	652.1	619.9	572.3	536.3
2.0	651.3	619.1	570.8	534.7
$P_p(P_{wf})$ (Psia)				
0 ($p_p(p_{ws})$)	5.093×10^7	5.093×10^7	5.093×10^7	5.015×10^7
0.5	4.379×10^7	3.970×10^7	3.403×10^7	2.979×10^7
1.0	4.352×10^7	3.922×10^7	3.348×10^7	2.936×10^7
1.5	4.332×10^7	3.911×10^7	3.330×10^7	2.919×10^7
2.0	4.321×10^7	3.901×10^7	3.312×10^7	2.902×10^7
Extended Flow Test Point		$P_{wf} = 567.7$ Psia, time = 24 h		
$Ps = 706.6$ Psia		$q = 2.665$ MMScf/d		
$Pp (P_{wf}) = 3.276 \times 10^8$ Psia^2/cP		$Pp = 14.65 = 2766.6$ Psia^2/cP		

Solution

1. Using Rawlins-Schellhardt analysis

By applying the Rawlins-Schellhardt method, Table 8.10 and Fig. 8.16 show the plotting function data points.

$$\Delta p_p = p_p(p_s) - p_p(p_{wf})$$
$$\Delta p_p = 5.093 \times 10^7 - 3.276 \times 10^7$$
$$= 1.817 \times 10^7 \, \text{Psia}^2/\text{cP}$$

Determine the deliverability exponent, n, for every line or isochron. Regression analysis with least squares should be used for this case. Deliverability exponents are shown in Table 8.11.

Table 8.10 Isochronal test data for Rawlins-Schellhardt analysis

$\Delta P_p \, (\text{Psia}^2/\text{cP})$

Time (h)	q = 1.520 MMScf/d	q = 2.041 MMScf/d	q = 2.688 MMScf/d	q = 3.122 MMScf/d
0	706.6	706.6	703.5	701.2
0.5	0.714×10^7	1.123×10^7	1.645×10^7	2.039×10^7
1.0	0.714×10^7	1.171×10^7	1.700×10^7	2.082×10^7
1.5	0.761×10^7	1.182×10^7	1.718×10^7	2.099×10^7
2.0	0.772×10^7	1.192×10^7	1.736×10^7	2.113×10^7

Fig. 8.16 Show the modified isochronal test data using Rawlings-Schellhardt

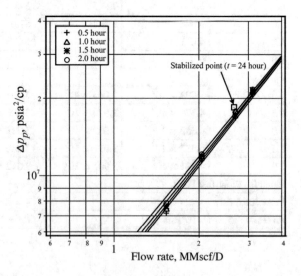

Table 8.11 Deliverability exponents for the modified isochronal test

Time (h)	Deliverability exponents (n)
0.5	0.72
1.0	0.74
1.5	0.74
2.0	0.78

Calculate the arithmetic average of Deliverability exponents (Table 8.10):

$$\bar{n} = \frac{n_1 + n_2 + n_3 + n_4}{4}$$

$$\bar{n} = \frac{0.72 + 0.74 + 0.74 + 0.78}{4} = 0.74$$

$$q = C\left[p_p(\bar{p}) - p_p(p_{wf})\right]^n$$

As $0.5 \leq n \leq 1.0$, Calculate the stabilized performance coefficient, C, using the values of the extended, flow test point and $n =$ average deliverability exponents (0.74). It should be noted that the pseudopressure used to determine the stabilized C value is assessed at p_s taken at the start of the test rather than p_{ws}.

$$C = \frac{q}{\left[p_p(p_s) - p_p(p_{wf})\right]^n}$$

$$C = \frac{2.665}{\left[5.093 \times 10^7 - 3.276 \times 10^7\right]^{0.74}} = 1.132 \times 10^{-5}$$

Calculate q_{AOF}:

$$q_{AOF} = C\left[p_p(p_s) - p_p(p_b)\right]^n$$

$$q_{AOF} = C\left[p_p(p_s) - p_p(p_b)\right]^n$$

$$q_{AOF} = 1.132 \times 10^{-5} \times \left[5.0935 \times 10^7 - 2766.6\right]^{0.74}$$

$$= 5.7\,\text{MMScf/d}$$

To estimate the AOF from the plot, draw a line of slope $1/\bar{n}$ through the stabilized flow test point, extrapolate the line to the flow rate at $\Delta p_p = p_p(ps) - p_p(p_b)$, and read the q_{AOF} (Fig. 8.17).

2. *Using Houpeurt analysis*

Plot the value of p_p/q versus q on a Cartesian plot (Fig. 8.18). The isochronal data points are shown in Table 8.12 select the best-fit lines using the modified isochronal data points for each iteration. The first data point of every isochron at the lowest rate does not fit along the same straight line as the last three rate points. Consequently, it's not accounted for in the calculations that followed.

Fig. 8.17 Show the
extrapolation line on
stabilized flow test point to
obtain q_{AOF}

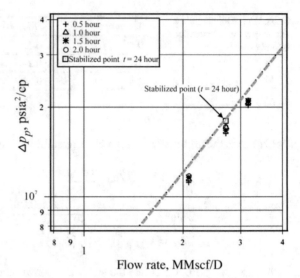

Fig. 8.18 Modified
isochronal test data using
Houpeurt analysis

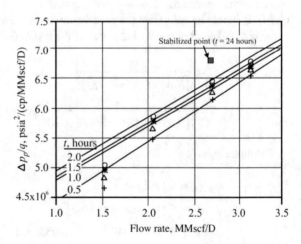

Table 8.12 Modified Isochronal test data using Houpeurt analysis

$\Delta P_p/q$ (Psia^2/cP/MMScf/d)				
Time (hrs)	q = 1.520 MMScf/d	q = 2.041 MMScf/d	q = 2.688 MMScf/d	q = 3.122 MMScf/d
0	0.470×10^7	0.550×10^7	0.612×10^7	0.653×10^7
0.5	0.488×10^7	0.574×10^7	0.632×10^7	0.667×10^7
1.0	0.501×10^7	0.579×10^7	0.639×10^7	0.672×10^7
1.5	0.508×10^7	0.584×10^7	0.646×10^7	0.678×10^7

Table 8.13 Slop "b" for the modified isochronal test of Houpeurt analysis

Time (h)	Psia^2/cp/(MMscf/D)^2 (b)
0.5	9.654×10^5
1.0	8.678×10^5
1.5	8.711×10^5
2.0	8.780×10^5

Fig. 8.19 Show the extrapolation line on stabilized flow test point

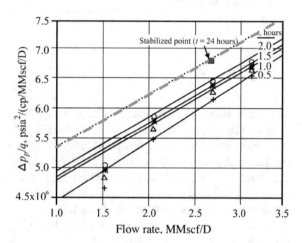

Estimate the slopes of each line, b, of the best-fit lines over the data points. Table 8.13 summarize the estimated average arithmetic slopes Fig. 8.19.

Calculate the arithmetic average of the slopes (Table 8.12):

$$\bar{b} = \frac{b_2 + b_3 + b_4}{4}$$

$$\bar{b} = \frac{(8.678 + 8.711 + 8.780) \times 10^5}{4}$$

$$= 8.723 \times 10^5 \, \text{Psia}^2/\text{cp}/(\text{MMscf/D})^2$$

Estimate the stabilized isochronal deliverability line intercept, a:

$$a = \left[\frac{p_p(p_s) - p_p(p_{wf})}{q} \right] - bq$$

$$a = \left[\frac{1.817 \times 10^5}{2.665} \right] - (8.732 \times 10^5) \times (2.665)$$

$$= 4.493 \times 10^6 \, \text{Psia}^2/\text{cp}/(\text{MMscf/D})^2$$

Finally, estimate the q_{AOF} using $\bar{b}$ and the stabilized a value:

$$q_{AOF} = \frac{-a + \sqrt{a^2 + 4b[p_p(\bar{p}) - p_p(p_b)]}}{2b}$$

$$q_{AOF} = \frac{-4.493 \times 10^6 + \sqrt{(4.493 \times 10^6)^2 + 4(8.723 \times 10^5)(5.093 \times 10^7 - 2766.6)}}{2(8.723 \times 10^4)}$$

$$= 5.5 \text{ MMScf/d}$$

8.7   Summary

Consistent values of compressibility and density are only taken into consideration within a certain pressure range when modeling liquid flow for the interpretation of well tests. So when the gas compressibility factor is also added for a more mathematically rigorous representation, like in the case of gas flow, this assumption is not correct. The liquid diffusivity solution can match the parameters of gas flow by normal linearization of the gas flow equation. Based on the viscosity-compressibility combination, three approaches for linearization are suggested: square of pressure squared, pseudopressure, or linear pressure. Assuming that wellbore storage conditions are negligible, drawdown tests are best analyzed using the pseudopressure function. Also, pseudotime represents the thermodynamics of the gas the best since the viscosity-compressibility combination is highly dynamic in the gas flow.

References

Al-Hussainy, R., Ramey, H.J. Jr., and Crawford, P.B. 1966. The flow of real gases through-porous media. Journal of Petroleum Technology, Transactions AIME. 18:624–636. *(PDF) Gas Well Testing*. Available from: https://www.researchgate.net/publication/318987018_Gas_Well_Testing [accessed Oct 18 2022]

Agarwal, Ram G. "Real gas pseudo-time-a new function for pressure buildup analysis of MHF gas wells." *SPE Annual Technical Conference and Exhibition*. OnePetro, 1979.. https://doi.org/10.2118/8279-MS

Ahmed, T. (2019). Gas well performance Handbook. Oxford: Gulf Professional Publishing.

Cullender, M.H. 1955. The Isochronal Performance Method of Determining the Flow Characteristics of Gas Wells. In Petroleum Transactions, 204, 137–142. AIME.

Katz, D.L. et al. 1959. Handbook of Natural Gas Engineering. New York City: McGraw-Hill Publishing Co.

Meunier, D. F., C. S. Kabir, and M. J. Wittmann. "Gas well test analysis: use of normalized pseudovariables." *SPE Formation Evaluation* 2.04 (1987): 629–636.

Rawlins, E.L. and Schellhardt, M.A. 1935. Backpressure Data on Natural Gas Wells and Their Application to Production Practices, Vol. 7. Monograph Series, USBM.

Schellhardt, Morris A., and E. L. Rawlins. Comparison of Output and Intake Characteristics of Natural-gas Wells in Texas Panhandle Field. Vol. 3303. US Department of the Interior, 1936.

Chapter 9
Practical Aspects of Well Test Interpretation

9.1 Factors Affecting Well Test Interpretation

The accuracy of the pressure and rate data utilized for analysis directly affects the interpretation's results. In well test interpretation, data preparation is essential and usually takes longer than analyzing well pressure responses. The common issues that arise while arranging data for analysis are addressed in detail in the chapter, along with data verification and validation. Additionally, will discuss how the well or reservoir conditions can affect the recorded pressure data. When test results deviate from typical wellbore storage behaviour, the identification of wellbore issues is highlighted as well.

9.1.1 Well Test Data Preparation and Verification

This Section uses the final build-up of the test pattern shown in Fig. 9.1 to demonstrate many probable errors may be accrued during data analysis. In this case, the well was pumped for 100 h, shut down for 50 h, and then reopened at the same flow rate for 20 h flow test before the final build-up. The reservoir performance throughout the test sequence corresponds to the infinite-acting radial flow regime.

9.1.1.1 Definition of Flow Rate History

Once the flow rate history data is prepared for a well test analysis, there may be two main challenges:

1. The production history for the well is incomplete or inaccurate. During several flow intervals of the test sequence, the flow rate must be determined.

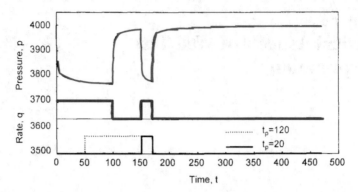

Fig. 9.1   Case of test sequence with two drawdown flow test periods

2. Numerous flow rate changes happened for the lengthy flow period before the test
   period of interest. Therefore, the flow rate history must be simplified.

A simple test scenario shown in Fig. 9.1 to demonstrate the effects of an erroneous
production history. The practical definition of the well production history applied for
interpretation is then presented.

9.1.1.2   Example of Rate Simplification

There are two ways for simplifying a test's rate history:

1. The ratio of cumulative production divided by the last rate is known as the
   comparable production time. The final build-up stage is analyzed in Fig. 9.1
   using a previous rate history simplified into a single drawdown of $t_p = 120$ h.
2. When there is a shut-in interval in the rate history and the bottom hole pressure
   has nearly reached the initial pressure, it is incorrectly believed that the rate
   history before this shut-in has no influence on the final build-up response and is
   neglected. With this example, tp = 20 h in the test case.

Figure 9.2 shows a logarithmic plot comparing the proper multiple rate derivative
response to the curves created by the two simplified rate sequences. When $tp = 20$ h
is applied, the superposition time correction is too long (Fig. 9.2 explanation), and
the derivative deviates above the theoretical stability corresponding to the radial flow.
With tp = 120 h, the intermediate shut-in from 100 to 150 h is neglected, the time
superposition function does not entirely correct the impact of the prior rate sequence,
and the derivative curve falls below the intermediate time stability.

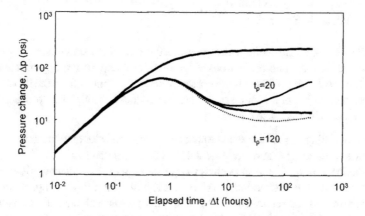

Fig. 9.2 Final build-up. The derivative is created with three different rate histories

9.1.1.3 The Concept of Rate History

When production changes took place a long time before the analysis period, it is practical to reduce the rate history; however, this is not practicable when rate fluctuations occurred just before the test period. The following rule of thumb is suggested by Bourdarot (1998).

Any rate changes that took place more than $2\Delta t$ before the period's commencement can be streamlined if the analysis period's length is Δt. The number of rate changes is then decreased using the comparable Homer time, only maintaining the most notable rate fluctuations and lengthy shut-in times. It is challenging to understand the model extrapolated pressure when the entire production period prior to shut-in is longer than the length of a build-up test. The extrapolated flow periods may be impacted by a variety of reservoir characteristics and boundary changes, even if they are not clear during the short build-up time test. In such a condition, the reservoir model obtained from the short build-up analysis may not extrapolate at very long periods.

When a lengthy production history is utilized in a multiple-rate sequence, the reservoir and boundary model must be relevant to the longest extrapolated period and contain all changes in reservoir properties and limitations throughout the vast investigation region of the first extrapolated period. If not, the model extrapolated pressure calculated from the build-up analysis is incorrect. When no separator measurements are taken during a clean-up or any other flow time, certain rate data are unavailable. The missing rates should be approximated before being added to the production history. However, if pressure data are obtainable during certain flow times, it is feasible to validate the estimated flow rate. Typically, wellhead pressure and choke size are utilized.

9.1.1.4   Periodic Error

After defining the rate history, the pressure data obtained downhole is split into discrete test periods, and the various values Δp, $\Delta p'$ and Δt are approximated for log–log analysis. All test periods are normally derived instantly from the rate changes stated in the rate history. Many errors may be produced through this procedure and affect the periodic response curves:

1. The beginning of the test period may occur earlier or later than the actual change in rate when the pressure and rate data are not fully matched.
2. Sometimes when a shut-in occurs, the pressure is erratic or fluctuating. For the beginning of the new period, $p(\Delta t = 0)$, the software utilizes the pressure point at the time of the rate change. The estimated pressure changes that result from this time might be either higher or lower than the actual stabilized pressure at the end of the prior period.

Figure 9.3 shows the shut-in time for the final buildup test shown in Fig. 9.1. There are five potential errors taken into account. Cases 1 and 2 depict time errors of 0.1 h before and after the shut-in time. For cases 3 and 4, a10 psi pressure errors below and above the final flowing pressure, and case 5, which corresponds to a time and pressure error, uses a buildup point at the beginning of the period.

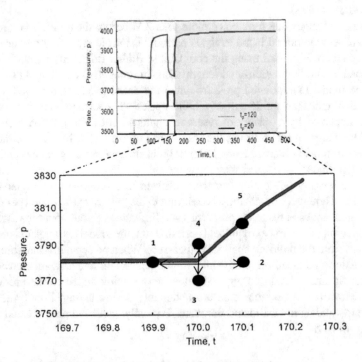

Fig. 9.3 Case of Fig. 9.1 at time of shut-in (time and pressure errors)

When there is a time error, the determined elapsed time Δt is either very larger (case 1) or very small (case 2). The pressure curve is shifted to the right in the first case, and it rises with a slope larger than unity (Fig. 9.4, case 1). As illustrated in Fig. 9.4 case-2, the pressure curve is shifted to the left when the shut-in time used to extract the test period is too late and distorted when it is applied too early. This error may indicate a linear flow regime at the early time if the quality of the pressure data is inadequate. It's interesting to note that the derivative curves are less substantially distorted than the pressure responses, allowing for early error detection.

Similar distortion on the pressure curves is seen when there are pressure errors. When Δp is overestimated (case-3, Fig. 9.4), the pressure curve is shifted upward, and at an early period, the distortion is quite similar to case-2. Case-4 d's response matches case-1 due to an underestimation of Δp (case-4, Fig. 9.4).

On a log–log scale, it might be challenging to see the error when a build-up point is utilized to start the period. In case-5, Δp, $\Delta p'$ and Δt were calculated at a certain point during the pure wellbore storage period (Fig. 9.5). The resulting pressure and derivative curves, which indicate that the result is accurate, initially follow a straight line with a unit slope. On such a test period, a favourable match can be found, but because Δp is too small, the resultant skin is underestimated. The distortion of the response is simpler to identify when the build-up point used to determine the start of the period is taken after the pure wellbore storage.

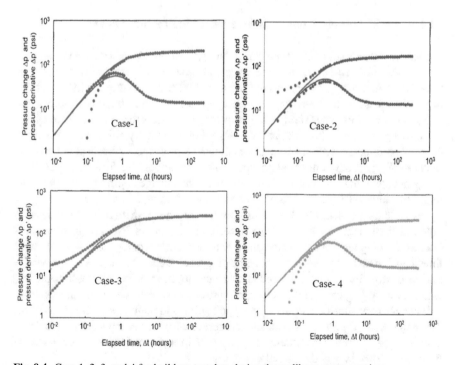

Fig. 9.4 Case 1, 2, 3, and 4 for build-up test data during the wellbore storage regime

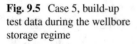

Fig. 9.5 Case 5, build-up
test data during the wellbore
storage regime

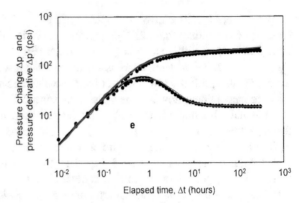

There are several techniques for correction when a log–log plot indicates a time
or pressure inaccuracy, as in Fig. 9.4. A linear scale plot of Δp versus Δt, can be used
to determine if the response is influenced by the wellbore storage effect. The WBS
slope (Mwbs) must intercept at the origin. Otherwise, the time or pressure correction
needs to be used, however, the linear scale graph doesn't show which parameter
needs to be altered.

9.1.1.5   Drift Measurements and Assessment in Pressure Gauges

A pressure gauge's suitability for a particular application depends on several factors.
Among the most crucial factors are gauge, absolute or differential, transducer or
transmitter, measuring range, fitting type and size, and absolute maximum ratings
like burst pressure. In these ways, a variety of gauges could be able to satisfy the
application requirements. Then, taking into account aspects that impact accuracy can
serve as a guide for making the best decision. Fundamentally, this establishes if the
pressure readings presented are accurate enough to guide the application's decisions.
Because of aging, environmental impacts, and other factors specific to an application,
gauge accuracy tends to deteriorate with time. The change coefficient of such drift
might be either positive or negative, and it is unpredictable.

Pressure gauge drift, also known as pressure transducer drift, is the slow deterio-
ration of these components, resulting in readings that are off from their initial calibration. Drifting leads the pressure gauge or transducer's accuracy to deteriorate with
time, causing incorrect pressure gauge readings. Each component of a gauge will be
made from a different material depending on the application (water pressure gauges,
water level gauges, etc.), and when exposed to certain conditions, these components
will change over time depending on the application and materials of the component.
When exposed to severe pressures and temperatures, gauge components expand and
contract, and they are also influenced by other environmental factors such as pressure
fluctuation frequency and material responses.

Pressure gauge drift influences the value of the pressure gauge as well as the reliability and accuracy of your readings. When seeking accurate data and data, pressure gauge accuracy is very crucial. Pressure gauges of all varieties are sensitive to drift/shift with time, regardless of the brand, cost, or reliability of the component. Components will expand and contract when exposed to high pressures and temperatures, and they will also be influenced by other environmental factors such as pressure change frequency and material responses.

Several pressure gauges are often run-down holes during the testing process to reduce the chance of a gauge drift. By calculating the difference Δp between the pressure signals, the gauge responses are compared before pressure transient analysis. If Δp is not constant, either one gauge is affected by drift, or the fluid column's weight between the two gauges is not constant.

Every pressure gauge will deviate from its output's calibration. Usually, the zero-point drifts, which causes the entire calibration curve to move upward or downward. A shift in the slope of the curve, represented by the span drift component, can also exist, as seen in Figs. 9.6 and 9.7.

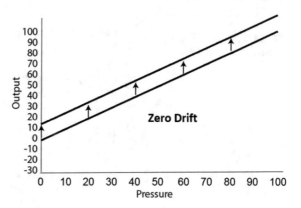

Fig. 9.6 Drift from zero point (calibrated output)

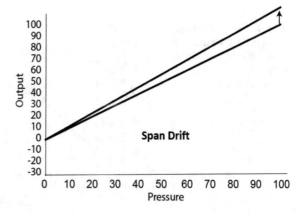

Fig. 9.7 Span drift

Drift can either be positive or negative, depending on whether the measured pressure is increased or decreased. In Fig. 9.8, a constant drift of 0.05 psi/hr is presented in the build-up example of Fig. 9.1. On the resulting logarithmic plot of Fig. 9.9, the derivative curves propose the existence of an apparent boundary effect, sealing in the case of positive drift, and constant pressure when it is negative drift.

The effect of a constant drift in reversed through flow and shut-in periods. For instance, a rise of derivative on build-up period is changed into a pressure steadying through drawdown test, so a dropping derivative curve. This can help to classify a problem of constant drift.

When there is only one pressure gauge accessible for analysis, the flow and shut-in durations on a normalized log–log plot (Ap/q and Ap'/q vs At) can be compared to identify pressure gauge drift. Gauge errors are possible when the responses are not symmetrical. A strong sign of a pressure drift can also be seen when looking at the test simulation on a linear scale. When a build-up response, for instance, demonstrates pressure stability in line with a dropping derivative response, for example on the negative drift curve of Fig. 9.10, a closed system or constant pressure boundary system can be seen.

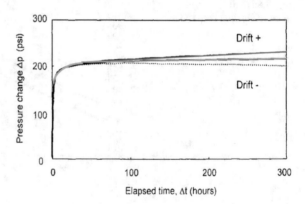

Fig. 9.8 Positive drift and negative on a linear scale

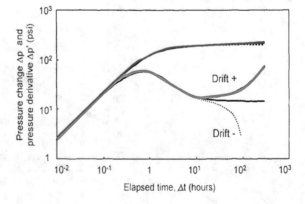

Fig. 9.9 Show the existence of an apparent boundary effect (positive and negative drift)

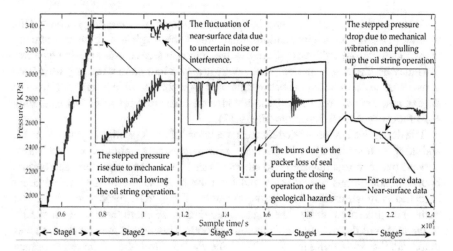

Fig. 9.10 Illustrate the noises during the five stages of well testing process

9.1.1.6 Noisy Pressure Gauge

When analyzing well test results, we may use downhole pressure data from permanent downhole gauges to determine various reservoir parameters, such as reservoir permeability, wellbore storage coefficient, skin factor, drainage radius, etc. The initial data from permanent downhole gauges typically contains significant quantities of noise for a variety of reasons, including gauge issues brought on by physical changes in the reservoir. The pre-processing of permanent downhole gauges data is necessary before further analysis since distorted data may result in a significant degree of error in the interpretations of well tests. Therefore, well tests results require the removal of noise. Osman and Stewart (1997) employed the Butterworth digital filter approach to remove noise from the data, however, Kikani and He (1998), demonstrated that it sometimes performed poorly.

Data processing will be more challenging since some stage curves have characteristics in common with noise. The features of test data from oil well testing are analyzed in Fig. 9.10. The entire well-testing operation procedure typically has five stages (Olsen et al. 2005). Although the near-surface data are simple to send, they are not as reliable as the far-surface data. Because there is no downhole operation during the waiting time, the far surface measurements are consistent, and the pressure curve exhibits minimal variation. In contrast, the near surface measurements in the same working stage fluctuate as a result of unknown noise or interference. The filtering procedure must specifically exclude these variations between near- and far-surface data. Actual downhole geological pressure signals that are combined with the sounds have properties that are comparable to those of a regular signal. Due to the changes in downhole geological conditions and the surface operation environment, burrs and stepped pressure also exists.

The smoothing algorithm may be used to minimize pressure gauge noise. This method performs well with random noise, but if the original data were manually manipulated before producing the derivative graphs, a regular noise might be added. This may be the case, if the pressure points are split into pairs with very little time between them, with each pair being separated by a considerable interval of time. In the example below, a regular noise is added to the final build-up of Fig. 9.1 by adding 1 psi for every 2 pressure points (Fig. 9.11).

The derivative response is widespread on the resultant logarithmic plot, Fig. 9.12, and after an hour it begins to fluctuate with increasing amplitude. The derivative curves appear to separate into two smooth branches as a consequence. The bottom branch is sometimes out of scale and the time of departure of the two apparent branches is much earlier than in the example of Fig. 9.12 (produced with a low number of pressure points). The logarithmic derivative curve then shows just 50% of the data, but the overall appearance is smooth. This layout may be deceptive as it appears that smoothing is not required and the upward trend of the upper branch may be misinterpreted for a reservoir response when it is just a truncated response. A stable match cannot be found because the logarithmic pressure response in this scenario does not support the derivative signature.

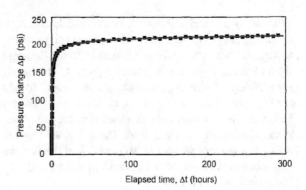

Fig. 9.11 Show the noise of plus 1 psi for each 2 points

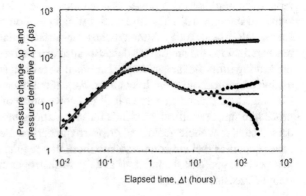

Fig. 9.12 Logarithmic plot shows noise of plus 1 psi every 2 points

It is helpful to highlight the derivative points on the logarithmic plot that are not depicted because they are negative or simply out of scale to avoid making a diagnosis on a truncated derivative curve. This is effectively achieved by colouring the missing derivative points on the scale's bottom x-axis with a different colour.

In general, noise on pressure measurements can be caused by a variety of sources, including pressure gauge errors (accuracy, resolution, drift), wellbore environment events (pump out, vibrations, etc.), reservoir events (tides, solid, etc.), or a mix of systematic and random components. Data interpretation becomes increasingly challenging as noise levels increase. Techniques for smoothing or characterizing errors might be used to reduce noise.

9.2 Impact of Well and Reservoir Conditions on Pressure Responses

9.2.1 Erratic Wellbore Storage

Typically, when one of the following circumstances prevails inside the wellbore, changing wellbore storage takes place:

(1) Changing the compressibility of the wellbore fluid.
(2) Redistribution of phases.
(3) Change in storage method from a liquid level that changes to a wellbore that is filled with liquid.

Phase redistribution happens in a well that is closed at the surface while gas and liquid flow concurrently into the conducting pipe. In these conditions, the effects of gravity lead the liquid to drop and the gas to flow to the surface. The redistribution of phases results in a net rise in borehole pressure since the liquid has very low compressibility and there isn't any additional space for gas to expand in a confined chamber. The additional pressure spike in the borehole is released via the formation when the phenomenon is observed in a buildup test. The formation pressure next to the borehole and the borehole pressure will eventually achieve equilibrium.

However, early in the process, the wellbore pressure could be higher than the formation pressure, resulting in an unusual hump in the buildup pressure that can't be properly analyzed by using only the dimensionless wellbore storage constant (C_D). Two models, presented by Fair (1981) and Hegeman et al. (1993), have been developed to address phase redistribution. These models incorporate apparent storage (C_{aD}) and the pressure parameter (C_{pD}), and two new dimensionless wellbore constants. The dimensionless anomaly pressure (ppD) rise's exponential form, according to Fair, is as follows:

$$p_{pD} = C_{pD} \left(1 - e^{-t_D/\alpha_D}\right) \qquad (9.1)$$

Later, Hegeman et al. demonstrated how buildup data with an unusual pressure reduction may be utilized to apply the negative C_{pD} values in the Fair model. As a result, they claimed that for certain wells, utilizing an error function to characterize the anomaly pressure may enable improved modeling of data sets with changing storage levels. Hegeman et al. thus suggested that:

$$p_{pD} = C_{pD} \cdot erf\left(\frac{t_D}{\alpha_D}\right) \tag{9.2}$$

In reality, both Fair and Hegeman et al. models are not significantly different. The impact of the three dimensionless storage settings on the dimensionless type curves is shown in Figs. 9.13 and 9.14.

During the early stages of drawdown, the pressure is high and no free gas is freed in the borehole. Initially, the response indicates the oil's compressibility. When the

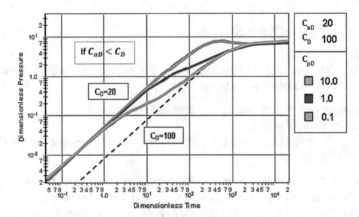

Fig. 9.13 Effects of three dimensionless storage parameters on the type curves (If $C_{aD} < C_D$)

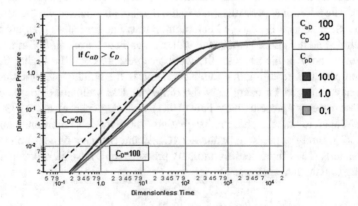

Fig. 9.14 Effects of three dimensionless storage parameters on the type curves (If $C_{aD} > C_D$)

Fig. 9.15 Shows changing wellbore storage coefficient for the drawdown test

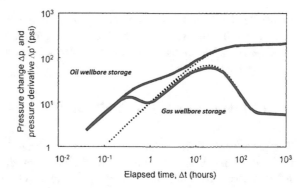

borehole pressure falls below the bubble point, gas compressibility takes over, and the wellbore storage coefficient is enhanced by changing Co to Cg. Figure 9.15 depicts a rise in the wellbore storage coefficient C on a logarithmic scale by a second unit slope straight line at later times. The pressure tends to stabilize and the derivative can display a slight falling trend during the transition between the oil compressibility wellbore storage Co and that for the gas Cg.

During the build-up test period, the response relates to the gas wellbore storage coefficient right after shut-in and then switches to the lower oil wellbore storage. This causes a sudden increase in derivative and, in some instances, a slope higher than unity by the end of the gas-dominated early time response, as shown in Fig. 9.16.

Due to the varied gas compressibility, when a significant drawdown is given to gas wells, altering wellbore storage can also be seen. The distortion on the pressure and derivative curves is less characteristic because the compressibility fluctuation is smoother than for oil wells below the bubble point. In comparison to theoretical models with steady wellbore storage, build-up data shows that the wellbore storage derivative peak is shorter. A constant wellbore storage interpretation model implies a lengthy derivative transition peak in such a situation where the early time unit slope straight line is appropriately matched, and it reaches the derivative stabilization later than the data. To effectively represent the start of radial flow at the start of the

Fig. 9.16 Shows changing wellbore storage coefficient for the buildup test

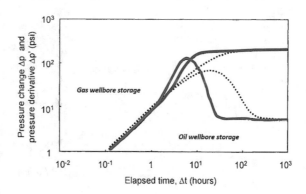

derivative stabilization, it is advisable to neglect the early time unit slope straight line and modify the wellbore storage coefficient on later elapsed time. Early on, the match is poor, but the reservoir response is well characterized.

9.2.2 Liquid Level at the Wellbore

The phase redistribution that occurs in the borehole during shut-in time causes a distinctive "humping" phenomenon for wells that produce multiphase flow such as oil, water, gas, or condensate. Here's a case of a well that is producing both water and oil to demonstrate this phenomenon. The gauge's depth shown in the Fig. 9.17 is higher than the formation. The weight of the fluid column between the pressure gauge and the sand-face changes as the water level rises when the well is closed and the water droplets fall to the bottom of the well. When the interface approaches the gauge depth, the hydrostatic weight eventually equates to 100% of water. As seen in Fig. 9.18, the build-up pressure often exhibits a small declining trend after a shut-in period.

The remaining build-up test data can be effectively evaluated after the interface between the two phases stabilizes or reaches the depth of the pressure gauge since, at that point, the pressure differential between the gauge and sand-safe becomes stable. When the build-up pressure is decreasing during the hump, the derivative turns negative as shown in Fig. 9.19.

The water cushion formed during the initial few hours of shut-in may be gradually pumped back into the reservoir at a later period. The build-up response may then be completely dominated by the effects of changing liquid levels, but only drawdown periods are relevant for evaluation (Gringarten 2000).

In general, it's best to place the pressure gauge as close as reasonably practicable to the perforation's interval. If phase redistribution is anticipated in a well that produces

Fig. 9.17 Hydrostatic level after shut-in and opening the well

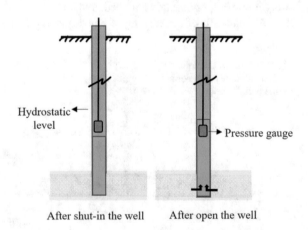

Hydrostatic level

Pressure gauge

After shut-in the well   After open the well

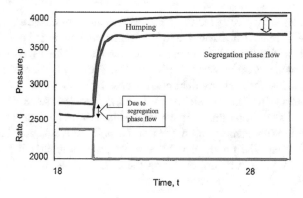

Fig. 9.18 Build-up test shows the Humping effect due to the phase segregation effect

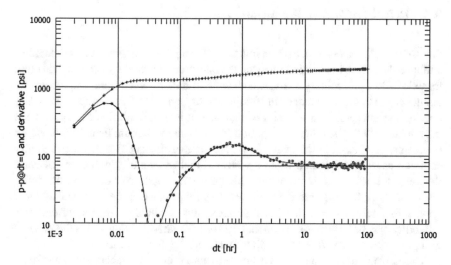

Fig. 9.19 Build-up test shows the phase segregation effect

many phases, the distance between the pressure gauge and the reservoir can be reduced to decrease the time the humping effect lasts.

9.2.3 Interference Effects

When testing wells in active fields, adjacent producer's well's interference effects may have an impact on the pressure data that is evaluated. A multiple well simulation models may preferably be used for the evaluation. The combined impact of nearby wells is added to the response of the tested well using the correct rate history for every producer and exact reservoir geometry. This process is time-consuming, and several estimates are typically required. For instance, it's possible that separate wells don't produce from the same zones, and it might be challenging to use an analytical model to characterize the well spacing and the geometry of the reservoir boundaries. A single well model is often used to evaluate tests. Therefore, it is advised to reduce as much as possible the pressure disruption caused by other wells.

The transitory effect is diminished as the time increases since the majority of well responses have a logarithmic time relationship. It is best to keep all other well's flowing conditions unchanged before the test if a well test is scheduled in a multiple well reservoir area. The potential interference impact is greater if a nearby well is opened or closed right before or during the well test than if its flow rate is left unchanged.

9.3 Effects of Static Parameters

Static parameter errors have a direct impact on the produced interpretation outcomes, although they often have little impact on the interpretation model selection. It is often feasible to do a preliminary analysis with approximations when static parameters are unknown and to further enhance the findings with modified values without materially altering the interpretation model. For instance, it is usually difficult to specify the net thickness "h". The range of variation for "h" can reach more than 30% based on the results of open-hole log analysis. A thickness error may also result from other common cases, for instance when a well is found to be partly perforated because the guns did not fire over the whole thickness of the reservoir or when the entire reservoir is not flowing into the borehole. In similarly, when the oil viscosity "μ_o" applied for analysis is calculated using correlations, the fluid parameter correlation's reliability may be quite poor. Given that the well test interpretation yields the "kh/μ_o" group, every error on "h" or "μ_o" directly affects the permeability "k".

The associated equations make it simple to determine how inaccuracy in one static parameter affects the interpretation results. Two factors are typically the subject of discussion:

Typically, the casing ID or the drilled hole diameter is used to establish the well-bore radius (r_w). The real wellbore radius is not properly specified, and the radial flow concept is not true near the perforations. The estimated skin factor is the only impacted parameter. The impact of any r_w, error is negligible, but for consistency reasons the same reference diameter must be applied when the aim of well testing is the change of skin during tests: the skin of a well should be specified concerning a selection of "r_w". Total compressibility "c_t" is determined by the saturations for each phase. This parameter is challenging to measure, especially in a reservoir at the bubble point pressure, when gas saturation varies.

9.4 Well Test Deconvolution

Deconvolution converts variable-rate pressure data into an initial drawdown at a constant rate throughout the length of the whole test, and it immediately produces the appropriate pressure derivative, normalized to a unit rate. This derivative is devoid of the inaccuracies provided by incomplete or shortened rate histories, as well as distortions brought by the pressure-derivative calculation algorithm (Everdingen and Hurst 1949). To choose and apply the proper pressure solution in the convolution integral, the convolution technique requires prior knowledge of the reservoir model. Superposition time is often calculated by convolving the reservoir rate with either logarithmic time or other suitable pressure solutions. The normalized/corrected pressure and the superposition time function are related linearly, and this relationship is used to estimate reservoir parameters (Kuchuk 2009; Gladfelter et al. 1955). The use of direct rate measurement in buildup test analysis was introduced by Gladfelter (1955).

The deconvolution process is an inverse issue. Without making any presumptions about the reservoir model and characteristics, this approach may produce the appropriate constant-rate pressure response. If there are no measurement errors, a direct algorithm can quickly and effectively handle the problem (Bostic 1980; Thompson and Reynolds 1986). But the direct approach is particularly unstable since it is susceptible to data noises. The direct technique is unable to yield results that can be interpreted, even with a modest degree of errors.

The solution has been subjected to several types of smoothness restrictions to minimize solution oscillation. However, the acquired results also are debatable, and the stability and interpretability remain unclear as the noise level rises. Deconvolution in the Laplace domain is an additional technique. Using the fact that deconvolution occurs due to division in the Laplace domain greatly simplifies deconvolution. Kuchuk et al. (1995) proposed a Laplace-transform-based approach for estimating the measured rate and pressure data using exponential and polynomial approximations, respectively.

In the absence of rate observations, the conventional constant-wellbore-storage model has often been employed to calculate the after-flow rate (Van Everdingen and Hurst 1949). In most real scenarios, however, the assumption of continual wellbore storage cannot be supported. According to Fair's research (1979) the wellbore-storage coefficient is not stable due to wellbore effects such as fluid compressibility, phase redistribution, and momentum impacts. The classical model cannot offer a meaningful explanation of the after-flow rate. In this case, normal deconvolution cannot be used, thus we must turn to blind deconvolution. Blind deconvolution is a method for deconvolving a measured signal without knowing the Kernel convolved Over the past 20 years, blind deconvolution has seen a lot of research activity in a variety of fields, including signal processing and image processing, but well testing has mostly avoided this field. Convolution kernel identification and signal response reconstruction must happen simultaneously in blind deconvolution.

Figure 9.21 is an example of deconvolution. The red curve in Fig. 9.21 is the deconvolved derivative derived by the deconvolution of the whole rate history in

Fig. 9.20. Its period is two orders of magnitude larger than the longest buildups, which are shown as distinct dots in Fig. 9.21. In Fig. 9.21, the change between the deconvolved derivative and the buildup data is due to the rate history before the corresponding buildups. After 10^4 h, the extended derivative revealed a contribution to production from a lower stratum. This was not visible in the longest buildups, which were restricted to 10^3 h.

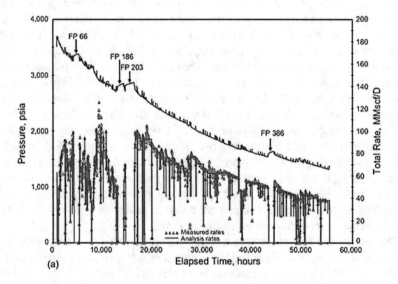

Fig. 9.20   Pressure and total rate history

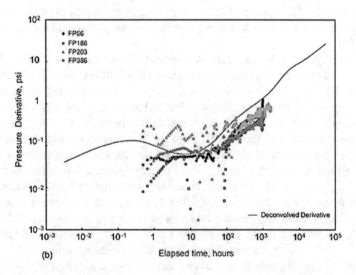

Fig. 9.21   Deconvolved derivative derived by deconvolution of the whole rate history

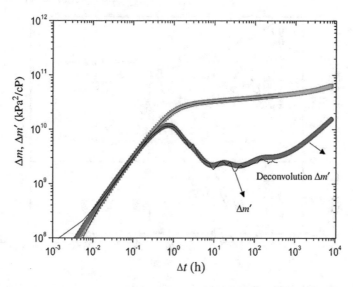

Fig. 9.22 Buildup analysis matching plot with the deconvolution method, (after, Jiang 2019)

Another example of the deconvolution approach is used to identify the border for the buildup test, as illustrated in Fig. 9.22. (Where the well presents the clear features of the closed boundary, which is interpreted to be 500 m). When the analytical well test analysis technique is inadequate for reservoir characterization, the numerical well test analysis approach, as illustrated in Fig. 9.23, may be used to characterize the reservoir by appropriately modifying the distribution characteristics of the reservoir region (Jiang 2019).

9.5 The Golden Rules of Accurate Test Interpretation

1. Well test interpretation entails far more than just pressure transient analysis.
2. Be extremely skeptical of all data.
3. Measuring anything is preferable to calculate it.
4. Accurate measurements are expensive. The bad ones are more expensive.
5. The longer the flow duration, the more information you can collect from any test.
6. While interpolation is typically safer than extrapolation, both should be employed with caution.
7. Whatever the gauge measures are not necessarily what the reservoir sees. Borehole transients would dominate reservoir transients.
8. If it occurs quickly, it is not reflecting the reservoir effect.
9. Examine the primary pressure derivative to distinguish between wellbore and reservoir influences.

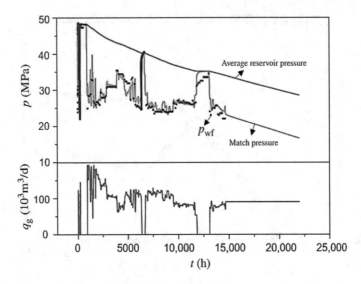

Fig. 9.23 Pressure history matching plot (after, Jiang 2019)

10. A buildup test for a single-phase fluid is continually concave downward in Cartesian coordinates. This is valid regardless of how complicated the reservoirs are, whether it is fractured, multi-layered (with equal pi), bounded, multi-permeability, or something else.

11. Show the position of the pressure gauge(s) and the producing zone on a wellbore diagram (s). Because pressure transient analysis uses with bottom-hole pressures, it could be necessary to convert wellhead pressures.

12. When it is practical, conduct a gradient survey just before and after a test.

13. Conduct a gradient survey as near to the wellbore's entry location as you can.

14. Carefully review the field reports. To ensure that the recorded measurements are accurate, for instance, compare wellhead values of temperature and pressure to those of the tubing and casing.

15. The temperature might disclose hidden information.

16. Always keep in mind that a computer is only a calculator, not an interpreter.

17. Maintain simplicity.

9.6 Summary

This chapter's last section discusses how the earlier theory may be put into practice. The test data may provide a variety of problems to start. The chapter covered in great detail how to prepare and validate the raw data that will be utilized in the analysis. Next, common distortions in the pressure response are discussed, followed by information on the variables that might alter how effectively tests should be interpreted. Also, the definition of flow rate history, the concept of rate history, period error, drift measurements and assessments in pressure gauges, noisy pressure gauges, and the influence of well and reservoir conditions on pressure responses are covered in the second section. The final section of the chapter addressed the erratic wellbore storage, impacts of static factors, the liquid level at the wellbore, interference effects, and well test deconvolution. The chapter concluded with the golden rules of accurate well test interpretation.

References

Bostic, J.N., Agarwal, R.G., and Carter, R.D.: "Combined Analysis of Post Fracturing Performance and Pressure Buildup Data for Evaluating an MHF Gas Well," JPT (Oct. 1980) 1711.

Bourdarot, G., 1998. Well Testing Interpretation Methods. Editions fechnip, Institut Franfais du Pdtrole.

Fair, W.B.: "Pressure Buildup Analysis with Wellbore Phase Redistribution," paper SPE 8206 presented at the 1979 Annual Technical Conference and Exhibition, Las Vegas, 23–26 September.

Fair Jr W.B., (April 1981). "Pressure Buildup Analysis with Wellbore Phase Redistribution", SPEJ 259–270.

Gladfelter, R.E., Tracy,G.W., and Wilsey, L.E.: "Selecting Wells Which Will Response to Production-Stimulation Treatment," Drill. and Prod. Prac., API (1955) 117

Gringarten, A. C., A1-Lamki, S., Daungkaew, S., Mott, R. and Whittle, T.M., 2000. Well Test Analysis in Gas-Condensate Reservoirs. S.P.E. paper 62920. Annual Fall Meeting, Dallas, Tex.

Hegeman, P.S., Hallford, D.L., Joseph, J.A. (September 1993). "Well-Test Analysis with Changing Wellbore Storage", SPEFE 201–207.

Jiang, Tongwen (2019). Dynamic Description Technology of Fractured Vuggy Carbonate Gas Reservoirs ‖ Well test analysis methods of fractured vuggy carbonate gas reservoirs. 61–133. https://doi.org/10.1016/B978-0-12-818324-3.00003-6

Kikani, J.; He, M. Multi-resolution analysis of long-term pressure transient data using wavelet methods. In Proceedings of the SPE Annual Technical Conference and Exhibition, New Orleans, LA, USA, 27–30 September 1998.

Kuchuk, F. J.: "Well Testing and Interpretation for Horizontal Wells," JPT (January 1995) 36.

Kuchuk, F. J. (2009). Radius of Investigation for Reserve Estimation from Pressure Transient Well Tests. SPE-120515-MS. https://doi.org/10.2118/120515-MS

Olsen, S.; Nordtvedt, J. Automatic filtering and monitoring of real-time reservoir and production data. In Proceedings of the SPE Annual Technical Conference and Exhibition, Dallas, TX, USA, 9–12 October 2005.

Osman, M.S.; Stewart, G. Pressure data filtering and horizontalwell test analysis case study. In Proceedings of the Middle East Oil Show and Conference, Paper SPE 37802, Manama, Bahrain, 15–18 March 1997. 2.

Van Everdingen, A.F. and Hurst, W.: "Application of the Laplace Transform to Flow Problems in Reservoirs," Trans, AIME (1949) 186, 305.

Van Everdingen, A.F. and Hurst, W.: "Application of the Laplace Transform to Flow Problems in Reservoirs," Trans, AIME (1949) 186, 305.

Thompson, L.G. and Reynolds, A.C.: "Analysis of Variable-Rate Well-Test Pressure Data Using Duhamel's Principle," SPEFE (Oct. 1986) 453.

Printed in the United States
by Baker & Taylor Publisher Services